# FRAGILE: HANDLE WITH CARE

### LIVING AND LOVING WITH AN EHLERS-DANLOS SYNDROME (EDS) DIAGNOSIS

## JULEE CRUZ

Difference Press

Washington DC, USA

Copyright © Julee Cruz, 2024

All rights reserved. No part of this book may be reproduced in any form without permission in writing from the author. Reviewers may quote brief passages in reviews.

Published 2024

DISCLAIMER

No part of this publication may be reproduced or transmitted in any form or by any means, mechanical or electronic, including photocopying or recording, or by any information storage and retrieval system, or transmitted by email without permission in writing from the author.

Neither the author nor the publisher assumes any responsibility for errors, omissions, or contrary interpretations of the subject matter herein. Any perceived slight of any individual or organization is purely unintentional.

This book is not intended to be a substitute for the medical advice of a licensed physician. The reader should consult with their doctor in any matters relating to their health.

Brand and product names are trademarks or registered trademarks of their respective owners.

Cover design: Jennifer Stimson

Editing: Madeline Kosten

*For my dad, for being there when I needed you the most.*
*For Gerad, Morgan, Logan, and Miri, who live this with me every day.*
*To Rebecca, Kim, and Christie, my genetically challenged sisters who*
*never fail to be there when I need them.*
*For Dr. Abu Aba Care, for saving my life more than once,*
*and giving me a hospital where I can feel safe,*
*where I know that in the future, I'll get the care I need.*
*In memory of Brandee, whose quick response to her intuition kept me alive (d. 2023).*

# CONTENTS

1. It's Not All in Your Head — 1
2. How I Earned My Stripes — 9
3. Finding Your Stripes — 23
4. Pain and Fatigue — 31
5. Living an EDS Life — 43
6. EDS Loves Company — 71
7. Finding the Right Doctor — 91
8. Prevention and Treatment — 111
9. Hospitalizations — 125
10. What's Really Going on in Your Head — 137
11. Support Is Vital — 147
12. Be Your Own Advocate — 167
13. Happiness Can Be Possible — 175

*Acknowledgments* — 191
*About the Author* — 195
*About Difference Press* — 197
*Other Books by Difference Press* — 201
*Gift for Reader* — 203

# 1

## IT'S NOT ALL IN YOUR HEAD

I still remember the moment I read my official diagnosis for Ehlers-Danlos Syndrome (EDS). I already knew what to expect. There were too many similarities; my life just all of a sudden made perfect sense. For all those years, everything seemed so random. There was nothing connecting all the illnesses, injuries, surgeries, and symptoms I'd had for so long.

Just one month before my fortieth birthday, I opened the letter and I read that, yes, I did have this life-changing genetic disorder, along with many comorbidity disorders and everything that came with it. I did have a very serious physical condition. And none of the symptoms were psychological in nature. I couldn't breathe. I thought that when I got my diagnosis that I'd feel like a huge burden had been lifted.

That finally, after twenty years of advocating for myself, having the answers to so many questions, and no longer being dismissed by doctors, that I'd finally be at peace. This was not the case.

As soon as I read those words, all I felt was dread. As a mother, I worried for my children's futures and was just downright terrified for whatever was coming my way. Yes, I got the answers I had been hunting for and the search was finally over. Time spent trying to find a diagnosis for my list of ailments could be over, since this was the underlying reason that my body wasn't working. What I was left with was the reality that I was never going to "get better," whatever that even meant, and that I would slowly decline as I aged. I knew what it took to get through an EDS life and what it took out of me to keep going, especially with no answers. The only comfort I got was that we had the opportunity to take precautions or prevention methods to ensure that my kids' futures were not as traumatic as mine had been for my family thus far. I had the chance to protect and keep their bodies in the best possible health, so they could enjoy a real life.

## PLUMBER

One of my favorite people, Kim, explains it like this: "You have a leak; the leak drives you crazy. You can't focus or sleep because all you hear is, *drip, drip, drip*. You call a plumber; he looks at it and tells you, 'There's no leak.' You pay the plumber. That night, you are kept up because you still hear the *drip, drip, drip*. You call another plumber to look at it, and this one says, 'I can't find a leak, I don't know what your problem is.' You pay that plumber." Now imagine repeating this 100 times, paying each time to hear, "There's no leak, it's all in your head." And pay I did – for twenty-three years, I reached the out-of-pocket cost with my insurance.

Before being diagnosed, there always seemed to be something going on with my health. If it wasn't a surgery (cyst removals, tonsillectomy, etc.), then it was an illness or injury (shingles, bronchitis, MRSA infections). Anytime anyone asked about me, there was always something medical going on; it just never seemed to end. I was in a doctor's office at least once or twice a week for a few years with whatever issue I was having at the time. What I do know is that medical professionals do not spend a lot of time studying genetics, with the exception of those pursuing genomics. There are so few doctors who understand connective tissue disorders and not all doctors have even heard about EDS. Another reason why so few doctors know about EDS is many who suffer from EDS have different experiences. Most disorders are easily identified with the combination of specific symptoms. Almost all symptoms of EDS overlap or mirror other disorders. If more doctors were aware of this disorder and how to screen for it, it would decrease the amount of time before getting a diagnosis. Since there is no cure for EDS, prevention and treatment of symptoms as early as possible will help make living with EDS easier and prevent symptoms from worsening.

WHY A ZEBRA?

In the medical field, doctors are taught to look for the most common answer. Most of the time, a patient has the illness that is presenting the most symptoms. Doctors are taught this concept with an analogy to horses. If you hear hooves, you automatically assume horses are making the sound. It's difficult to tell, if not impossible, to separate a zebra – the EDS society mascot – from the herd, due to the sound of the hooves, or in our case, our symptoms. EDS patients are considered zebras because our bodies are seen as normal. Most tests will come back normal.

Since EDS has so many symptoms that resemble other disorders and most people with EDS experience the syndrome differently, it is difficult to diagnose someone with EDS. This combination of illnesses always seems random and unconnected to each other, so EDS is easily overlooked. As time went on for me, it didn't seem likely that I exhibited fifty or more different disorders and symptoms, like diverticulitis, pancreatitis, gastroparesis, or severe IBS, to name a few. It was more likely that there was one all-encompassing disorder that caused all these other disorders.

## IT'S CELLULAR

For all observations or tests, it is not possible to look at the cellular level of everything that's going wrong with your body at any given moment. To describe EDS in a simple way, imagine a body without EDS. For these people, their skin acts as a rubber band. Pull it back a bit and let go, and it snaps right back into place. For people with EDS, our body works more like a chewed piece of gum. If you pull a piece of chewed gum apart, it will stretch out instead of snapping back together like a rubber band.

EDS is a connective tissue disorder, so in this example, the skin doesn't bounce back as easily once it's stretched out. It is much more difficult, if not impossible, for it to return to its former shape. At the cellular level, EDS is mutated or abnormal collagen. Collagen makes up 70 percent of the protein in your skin and as much as 30 percent of protein in the body. This means it affects every organ, all fluids, and every system. This weakens all the tissues in your body. It causes all joints to be more lax and more susceptible to dislocations. Since the ligaments and tendons are so lax, they are unable to keep our bodies held together alone. It's our muscles that end up keeping our joints and bodies held together, causing a lot of pain.

THE DIAGNOSIS IMPACT

Once I found out that I was different from everyone else, the expectations to be normal were gone. I no longer pushed myself beyond my limitations, no longer forced myself to find energy I didn't have. For years, I kept enrolling to college and was working towards a bachelor's degree in psychology – just to be forced to drop out repeatedly. My diagnosis helped me learn that it wasn't about me; I wasn't stupid, or incompetent, and dropping out wasn't a personal failure. I knew that the degree I had wanted for so long was no longer attainable and the hardest part was that there was nothing I could do to change that.

I had a condition that demanded certain medications that affected my brain, making it impossible to retain information. Having EDS and taking medications didn't mean that I couldn't make something of myself. I could still live a successful life full of wonderful experiences, full of love, and certainly full of fun. I only had to accept that my life was also going to be full of hospitals, doctors, nurses, medications, and pain.

However, learning about EDS and making my life make sense still didn't hold the weight of what I felt when I got the official diagnosis. Before, there was always the hope of getting better or making a full recovery. My diagnosis hit me hard. I had doctors who are well established in their careers with at least a decade or longer of experience tell me that I've been their sickest patient, that I have the weirdest anatomy, or that I have things that they've never dealt with before. Some say I have issues they wouldn't believe possible only if they hadn't seen it with their own eyes.

Maybe it wouldn't have hit so hard if it was only me who would continue to have health issues. Having been so sick in my twenties and thirties, I can't help but wonder what will happen next as I get older. But even all of that doesn't hold a candle to what I feel for my children, Logan and Morgan, and their future. I know my kids are resilient and incredibly strong. At some point, EDS will show up. There's nothing I can do about that, but I do everything I can for my health. I want to be here as long as I can for my children.

I'm glad I never stopped looking for answers. I never want my kids to have to hear that it's all in their heads; they saw it happen to me. Writing this book is my way of fighting back, my way of helping and spreading awareness to help others struggling with EDS, and my way of one day, putting a stop to anyone else being told it's in their heads.

## 2

# HOW I EARNED MY STRIPES

Strong, courageous, resilient, brave, a warrior, even a hero – there are so many characteristics people have used to describe me. The truth is, I've never felt any of those things, especially during the worst experiences of my life. When those moments happened, I felt the weakest. I felt anxious, beaten, vulnerable, and scared. Most of the time, I felt I was pulled in too many directions all at the same time. I was completely falling apart and there was no way out. I was left with one question: what do I do when I don't really have a choice? I decided there was only one thing that I had control of and that was me.

EDS has taken away my ability to get a doctorate, drive my car, swim in a lake, hike in the mountains, ride a bike, be alone, and my ability to eat. So, then I decided *f\*\*\* you,* EDS, I won't let you take me too! I may not be able to do all these things anymore and I really didn't want my kids to grow up and only remember me being miserable. That was my first thought because I grew up in a miserable home. You'll struggle with all of this too, and we all have our good days and bad days.

To combat this, I decided to write the name of everyone I love and who loves me too. I started with my husband, my children, Morgan and Logan. I wrote down the kids we added along the way, Fei, Eva, Rachi, and Miri, and my new relationship with my dad, whom I've decided to be like when I grow up. He's had a very difficult life, and he didn't let it take away who he was. He is kind, compassionate, and he loves unconditionally. He's always there when you need him. He works so hard while making others around him feel good about themselves, smile, and laugh with his great sense of humor. The best thing about him is his limitless generosity. He is insanely genuine and thoughtful toward those he loves, even if it's something as simple as obsessing over buying the perfect coat for someone.

After that, if you want, you can make a list of everything you can still do despite your diagnosis, whether alone or with help. When you're done with that, and nothing pops up, then at the very least there are some silver linings. Remember that your legs are as soft as baby's butt, and we will end up with the fewest wrinkles.

## BE STRONG AND RESILIENT

Too often we don't see ourselves as others see us. I'm not sure if I'll ever feel brave or strong, even though I see it in all my friends who have EDS and face a lot of the same things that I do. We all struggle with the feelings that our lives are pointless and that we have no real purpose. Our lives revolve around EDS and the fact that it takes a lot of our time to keep up on the treatments and therapies that keep our bodies functional and feeling as good as possible. The worst part in having EDS is that it's very isolating. There are a lot of activities, social gatherings, and personal dates that we must decline or cancel last minute. Everything depends on how we feel, and there's no way to know that until the morning of. We can try to help prepare our bodies by resting for a day or two prior to the event and resting a few days after so that our body can recover. Even then, there are many times that our bodies can become ill, have an injury, or when the exhaustion and pain are too much. It's impossible to keep up with everyone, let alone our own lives.

Normally I keep things about my health oversimplified or vague. In my experience, people only ask how I'm doing to check in but are not really interested in what is going on with me because everything has been very complicated. It's difficult to understand, and it usually takes longer to tell them than they expected when they asked.

For me, EDS started when I got pregnant with my kiddos. I had severe stroke symptoms during my first pregnancy, symptoms that I eventually found out were called transient ischemic attacks (TIAs). At the time, I was only twenty and I was very naïve. I didn't realize or recognize that there was anything wrong. At least, I didn't think it was anything severe enough to warrant any medications or testing that could cause distress to me or my unborn daughter.

My second child, my little boy, has been kind of a miracle to me. I was caring for two little girls along with my own daughter and started to have a TIA. I was at the store and had the three very young children with me when I started to feel numbness on my left arm. I tested to see if I was having a hard time with my words by talking out loud to myself. I stumbled on a few words, and it wasn't long until my vision became blurry. With my vision only getting worse, I didn't trust it enough to drive home. We walked around that store for about three hours until my vision returned to normal.

But still, this scared me. *How was I supposed to watch over three toddlers in this condition?* This is what led me to seek medical intervention. In trying to find the cause of all the TIAs, I was on multiple medications and underwent numerous tests that could be harmful to the baby. I gained 100 pounds in six months. After Logan was born, I lost all the weight and had a few of my healthiest years. I mentioned my son was a miracle to me – a miracle because even due to all the tests and medication, he was born without any adverse effects.

Since then, I've had thirty surgeries, twenty-plus within the last fifteen years, and three times I've had two surgeries back-to-back. I've had six surgeries on my wrists, three bladder surgeries, and two ectopic pregnancies (six weeks apart and one of which led to my tube bursting). After the ectopic pregnancies, I learned that my body couldn't carry another child; the muscles holding my bladder were stretched so much, I was told it looked like Swiss cheese. I've also had a sinus surgery after a nine-month long sinus infection, and a tendon repair on my ankle that had me in a boot for almost a year. I've had my tonsils and gallbladder taken out, a hysterectomy, and only one of my hernias repaired. Later, I had another total "Roto-Rooter" of the entirety of my chest cavity, with doctors scraping the sticky, tacky film off the inside of my chest cavity and off my organs. I had titanium rods inserted to replace vertebrae that were eviscerated and eaten away by osteomyelitis, and an extra rod to keep me upright because of the scoliosis I acquired. I had an extremely expensive and totally unnecessary bariatric sleeve that I had spent a year jumping through hoops to get. Finally, the most terrifying and life changing surgeries were open heart for a birth defect I didn't know I had.

So, that's surgeries – on to other random things! I was hospitalized for pancreatitis and ischemic colitis, I have had fifty plus outbreaks of shingles on my face, and I have diverticulosis which flares up on occasion. My neck, hips, and lower back are affected with degenerative disease which I get injections every three months for, and I've had an abscess that turned out positive for MRSA. Then I've had many bouts of bronchitis, colds lasting three to four weeks, and anything else my kids brought home from school. There was a year I literally would Lysol my kids from the neck down in the school parking lot before they were allowed to get in the car and then immediately passed around the hand sanitizer.

I've had battles with autoimmune diseases for who knows how long before I found treatment. I was diagnosed with Sjogren's and Reynaud. After finding my autoimmune doctor in Blackfoot, Idaho, 164 miles and an almost three-hour drive from my home, my body had shut down completely. According to this doctor, I was months away from lupus and rheumatoid arthritis and less than five years away from Crohn's disease, ulcerative colitis, or colon cancer. I had to drive weekly to Blackfoot for IV treatments for nine months and tapered down to once every six weeks. Along with autoimmune diseases comes delayed-onset reaction allergies, which was manageable until 2013, when I began having severe pain every time I ate or drank. This was all before being diagnosed with EDS.

Since being diagnosed with EDS in November 2019, I've gone septic twelve times and had three bowel bleeds. I've been in several comas, had two surgeries, spent months in physical therapy, spent two and a half years out of the last four years inpatient at hospitals, and came very close to death seven times in one year. After going septic my first time in 2019, I started passing out multiple times a day, some days as many as ten times. It can take anywhere from one minute to four hours before I wake up all the way. After I pass out, my heart doesn't go into tachycardia to help redistribute my blood to wake up. I have had to use a PICC line, which is an IV put in your arm that ends next to your heart. Due to IV potassium, my veins are very difficult to access. Also, with a PICC line, I was able to give myself IV fluids daily at home which prevented me from passing out. It was very difficult on all of us; our kids, nurses, and even the doctors were left blubbering when they saw me passing out so much.

Passing out constantly is awful, not knowing how long it will be before I wake up, although I can still hear immediately afterwards. It's exhausting and you are confused for a little while after you come to. I have hit my head on walls and floors. I've passed out once on the stairs, falling down the last three steps, and landing on my tailbone very hard before hitting the tile floor. I was sore for weeks. I've hit my head on almost everything and a few times I have really bruised and bloodied my face. While the physical struggles vary from person to person with EDS, one similarity is our comorbidities, which you'll read more about in a later chapter. I've suffered from mitochondrial disease and dysautonomia, including POTs and MCAS.

No matter how much you feel like you've gotten a handle on your life, after taking years to accept what's happened to you and knowing that something is always just around the corner, nothing will ever prepare you for it. I'd gotten used to the fact that I had been extremely close to dying so many times. Some of those times I still can't believe I survived.

There will be times when you'll feel like you have the wind knocked out of you and the rug pulled out from under your feet at the same time. I had made it a year without a major health crisis and for the first time in over a decade, I was hopeful about getting some of my life back. That's when my good friend EDS reminded me that there are no guarantees and the possibility that there may never be another calm before the storm.

For about three weeks, I was experiencing very intense pain around my right shoulder and nothing I tried gave me any relief. Afterwards, it was clear that this was the same intensity of pain that I had experienced when my spine had been eaten away by osteomyelitis. The pain was on the right side of my chest and had developed a hard spot under my skin. It started around my clavicle and went up to the middle of my neck. I went to urgent care to have it looked at and was given an X-ray. The result showed that I had pneumonia again so, with my history with pneumonia, we knew at some point we'd end up at the hospital.

A few hours later, the pain was too much and I decided to go to the hospital. I had blood cultures taken and a CT scan of my chest. Then pneumonia became the least of my worries, because this test showed that there was a huge abscess that tunneled under my ribs to my lungs. I was transferred to another local hospital by ambulance to see a thoracic surgeon to remove the huge ball of infection. The next day I went in for surgery and I prayed that they didn't have to open my chest entirely. What was discovered once they opened me up, was that I had osteomyelitis again and it had eaten through my first right rib and a good portion of my clavicle. I was left with a huge hole in my chest, this one a huge open hole. It was about 9 by 4 cm and it was 7.5 cm in depth, roughly the size of a baseball!

I had to have had this infection of osteomyelitis for six or more months without knowing about it. After surgery, I learned that the did remove my first right rib and cut off a portion of my clavicle. If anyone has had a large wound, they are familiar with this, and I had a wound vac put on after a few days after surgery. This is pain like never before; first, the wound is washed, then a spongy foam fills the entire wound. In earlier times, leeches were placed around the area to help bring the blood to the area to allow it to heal. Now they use a small machine to have that negative pressure bring the blood to the surface. In the beginning, I had to have it all taken off, cleaned, and patched back up three days a week. The worst part is that I have mast cell reactions to even the sensitive skin dressing, so I'm super sensitive there.

The process is the worst thing to have to endure for three days a week for months. Lidocaine is not used most of the time. You could have the dressing changed faster than it would be for lidocaine injections to work. In the beginning, changing my wound dressing was so painful that I passed out several times.

I was discharged about a week later, and within the next week, one night I had become unresponsive, and my husband couldn't get a blood pressure reading from me. Then I was taken again by ambulance back to the hospital. The doctors told us that had Gerad not called the ambulance, I would not have lived through the night. I was in complete and total renal failure and my skin was very jaundice looking. This time is just a blur to me dealing with so much trauma and unimaginable pain. The other side of my neck had tubes in the bigger vessels by the heart and I had to have a full dialysis cycle every day for about three hours until the antibiotic vanco was completely removed from my blood. Essentially, it was having so much of the antibiotic levels in my blood that had put my kidneys in acute failure.

I was able to come back home one to two weeks later. The next day, I was to go to a hospital for a two-hour transfusion of a newer antibiotic and get this IV therapy, and this ended up being a very scary experience. Within thirty seconds, I knew something was very wrong and I was barely able to tell them to stop the IV before I passed out. I wasn't out long and when I came to, I had felt deathly ill. I was bright red, and my throat closed up. I was wheezing trying to get oxygen. Now, I've had mast cell and allergy responses in many different ways, but never had I experienced a serious anaphylactic reaction. This time I had severe anaphylactic shock and within a minute I was given IV Benadryl, an IV steroid, and jabbed with an epi pen. In no time, I was put in a chair and the nurse pushing me ran as fast as she could to the ER. Within a few minutes, I was already in a room and getting more Benadryl, steroid, a full syringe of adrenalin, and finally a breathing treatment before I was able to breathe normally.

All of this was within a two-week period – a major surgery removing bones leaving a huge hole in my chest and two more times being in a life-threatening situation.

It's going on three months with a few more to go before I'm completely healed. Now I have three tubes attached to some kind of machines and need to take all of them every time I leave the house.

Through all of these trials, I've earned my stripes. I learned that being brave isn't actually feeling brave. You're scared, terrified, and lack confidence. Sometimes you experience low self-esteem. Being brave is when you come face-to-face with death and impossible odds, or any other type of situation of fight, flight, or freeze. We will fight 100 percent of the time, and for many of us with EDS, that fight is the only reason we are still alive.

# FINDING YOUR STRIPES

Any doctor can diagnose EDS. The most common type of EDS is hypermobility EDS (hEDS). The criteria for hEDS is:

"The clinical diagnosis of hEDS needs the simultaneous presence of criteria 1 and 2 and 3. This is a complex set of criteria, and there is much more detail than presented in this overview; please see the page for hypermobile EDS.
Generalized joint hypermobility (GJH) and two or more of the following features must be present (A & B, A & C, B & C, or A & B & C):
Feature A – systemic manifestations of a more generalized connective tissue disorder (a total of five out of twelve must be present)
Feature B – positive family history, with one or more

first degree relatives independently meeting the current diagnostic criteria for hEDS

Feature C – musculoskeletal complications (must have at least one of three); and

all these prerequisites must be met: absence of unusual skin fragility, exclusion of other heritable and acquired connective tissue disorders including autoimmune rheumatologic conditions, and exclusion of alternative diagnoses that may also include joint hypermobility by means of hypotonia and / or connective tissue laxity." (EDS Type, 2024)

## GETTING MY DIAGNOSIS

Those of us with EDS are normally not properly diagnosed for more than five to ten years for men or fifteen to twenty years for women. When I first learned of EDS, I took the information to my primary care provider. He had been treating me for over the last fifteen years and had seen firsthand everything that I had gone through. He told me that it made more sense that I had one underlying disorder that caused the seemingly unrelated health issues rather than having eighty different comorbidities and symptoms. I was never directed to genetics or suggested to get any type of testing done from any doctor that I had seen, and I've seen over 100 doctors. The way I got diagnosed was very odd.

My twenty-year high school reunion was coming up. I hadn't attended any previous reunions. I had graduated a year early and never looked back. I had a few old friends contact me and ask how I was doing. I responded and let everyone know, telling them a long story about what I'd been through. When I laid it all out, it seemed like a complete medical circus in chaos, complete with surgeries, autoimmune disorders, injuries, and illnesses. It caught the attention of a girl who I hadn't known while at school, Shelly. She told me a friend of hers had this disorder called EDS and based on what she had heard from my snippet, she thought I might want to investigate EDS. I read up on EDS, and when looking back on my life, everything made sense! All the unrelated illnesses and injuries that seemed random started to make sense in this context – my gallbladder problems, the many cysts, other autoimmune disorders, my inability to eat without pain, and my shingles outbreaks. All of these things didn't seem to relate to each other, but with what Shelly was suggesting, maybe there was a common denominator.

## BEIGHTON SCORING SYSTEM

I started researching more and I came across a simple way that I could test myself to make sure I was on the right track. The "Beighton Scoring System" is a diagnostic test you can do on your own to measure your degree of flexibility, and there is such a thing as too flexible, which is called hypermobility. Many people who can perform the extremes on the "Beighton Scale" are said to be "double jointed," but really, their ligaments are just not holding their joints together as tightly as they should.

This quick test shows a few things that you can do with your hands, or by bending over at the waist and touching the ground without bending at the knees. It takes two to three minutes to go through the nine things that they ask for and, depending on how many of those you can do, makes it easy to determine if you should look more into hypermobility as a problem rather than just considering yourself double jointed. I had been seeing a doctor for getting more cysts removed from my hands. Since he was also an orthopedic surgeon, I discussed with him the possibility of hypermobility and within minutes he determined that I had a nine out of a nine, and I had something else going on here, other than just having a few cysts removed, getting a cold or flu, and the occasional injury.

## FIND A GENETICIST

Once I connected this information, I began investigating the local geneticist for a diagnosis. I was told that his patients were on a waiting list because he was scheduling appointments out at least three years. Three years!? I didn't want to wait three years for a diagnosis. I wanted it much sooner. This would give proof to other doctors that I had the disorder – that I was not pretending or faking my symptoms.

I joined a local EDS support group and started asking questions. I found out that there were doctors out there who were willing to work with patients remotely. At the time, there were two doctors I learned about, one in Texas and another in Florida. I opted for the doctor in Texas simply because he charged less than the doctor in Florida. Both doctors had a cash price and didn't work with insurance companies. I contacted the doctor in Texas and was given very extensive and detailed paperwork that I needed to fill out and send back to him.

By the time I had completed the paperwork, I had at least eighty comorbidities / symptoms that were related to EDS including my history of thirty surgeries and my physical flexibility. I ended up being one of the worst cases the doctor had worked with over his fifty years in practice. After sending in the completed work, I had a clinical diagnosis within two weeks. At the time, I was diagnosed with EDS, POTS, MCAS, dysautonomia, and others.

The next step I took was to send in DNA samples to determine which of the thirteen types of EDS I had. Like I mentioned, most people are diagnosed with hEDS. This type does not currently have any genetic markers that have been found yet and relies on the clinical diagnosis. Some of the more serious types of EDS have known genetic markers. My test results came back with something a bit rarer than just being passed down genetically. Both of my biological parents had the same genetic marker out of the tens of thousands that were tested, where they received one normal gene and one mutated gene. My geneticist indicated that I had received the mutated gene from each parent. The gene affected is one that is known to have early onset Parkinson's disease and also affects collagen production. The geneticist diagnosed that I had myopathic EDS, the type that affects the nerves and muscles in your body. I had a history of problems with both nerve and muscle. Had I been tested and known earlier in my life, I would have taken different care of my body, as there is no cure for EDS and some of the best things for EDS patients are treatments for symptoms and prevention.

For most of us, a diagnosis gives us answers and a place to start, and a path to follow to get the right treatment. It gives us a meandering path to heal, a way to live life better. Unfortunately, our diagnosis gives us context for our lives and more questions than answers. Having a condition that doesn't have a direct path to follow is a very difficult thing to accept. Not knowing where to go and what comes next is terrifying, just like being dropped off in the middle of the desert and seeing nothing all around. We all have to find out what combination of treatments works for each of us. This book won't be able to give you all the answers you need, but this book is something I wish I'd had when I was diagnosed. It's not a complete map, but it will give you many starting points, a better understanding of EDS, and a better understanding of yourself.

# 4

## PAIN AND FATIGUE

I had always been an active person. I loved hiking in the mountains, camping, and swimming in lakes and beaches. When I really got sick, any movement came with pain and the fatigue was relentless, only making my body ache even more. My muscles atrophy easily due to the type of EDS I have, making it painful to lift, carry, drag, and push anything. Coupled with poor proprioception (the ability to sense the location of our limbs in relation to our bodies), and no balance, this makes me super clumsy, causing collisions and "body benders" like hitting walls, doorknobs, and pretty much anything that doesn't move. I have endless, tender bruises and my joints are stiff.

Pain also comes from our tendons and ligaments being weak. Since our muscles hold us together, and muscles are not meant to do this, the result is a lot of muscle pain and stiff joints.

All of these symptoms changed me, and I desperately missed my life and who I was before. Coming to accept that is one of the biggest hurdles when diagnosed with EDS. Life is never the same and we must change our perspective on everything quickly, as things change moment-to-moment with a surgery, illness, injury, or hospitalization. I had to learn to adjust to my limitations and still find a way to still feel like myself.

## GRIEVE YOUR LOSS

It's important to allow yourself to grieve this loss of function, loss of health, loss of the life you lived, and most importantly, for the loss of yourself. At some point, we have all lost someone who we loved and had to learn to cope with our loss as time goes on. In the same way, you need to allow yourself the time to grieve. In time, you'll be able to find your new self and how to adjust and live your new life.

Take the time to cope a little at a time. Though the loss can be sudden, learning to live afterward is a long process, both mentally and physically.

We must allow ourselves time to grieve for our own lives of who we were and of where we were going. All the hopes and dreams we had planned on achieving and all the life experiences we wanted to experience abruptly changed. Being diagnosed with a disorder that will change every aspect of your life can easily crush you and take all your life from you if you let it. It would be so easy to let depression take over or to become a recluse. It is easy to spend every day in bed with a T.V., laptop, book, or phone to keep you company, rarely speaking and sleeping it all away.

Just because we can't have the life we planned doesn't mean that we no longer have options. There is still life to be lived, people to love, and new things to experience. I had to find a way to transition, a way to allow myself to be happy, to succeed, and to be surprised again.

When diagnosed or in pain, we feel numb, like we are just going through the motions of life rather than living and feeling alive. We need to allow ourselves the time to accept the situation a little at a time, because if we don't, then that proverbial elephant in the room will surely just sit right down on top of us and squash our ability to work through our grief. As we continue to accept our situation, we begin the healing process.

## DON'T ALLOW EDS TO DETERMINE WHAT COMES NEXT

Allowing ourselves time to grieve is only the beginning. Learning how to live our new life doesn't have to be hard. The next thing we can do is to learn how our bodies work and how to manage them. Understanding our bodies can allow us to adjust and create a new life that still has goals, dreams, and happiness. I found that my two most difficult changes were the fatigue and pain. Being able to manage an EDS body will allow us to keep our pain and fatigue levels low.

We need to redefine how we view success. What we previously thought of as success will most likely become an unattainable goal. There is no keeping up with the Joneses. Our lives will switch gears and we shouldn't continue to race with healthy-bodied people. We can't keep up and compete; if we try, not only will we fail, but we will succeed at being miserable. The harder we try, the worse we'll feel. It won't take long before your body will quit working. It just isn't possible anymore. Once your body stops, your mental health would be next to decline. The longer you wait in accepting your EDS, the harder it will be when you decide that you don't want EDS to win.

We are all so much more than our diagnosis. We can still reach goals, enjoy success, and allow ourselves to be happy. Deciding to accept your EDS diagnosis and to alter your perspective gives you the power back and gives you a sense of control over your life. By taking the control back, you are opening yourself up to still love, to experience accomplishment, and feel joy. By accepting that we are different, the pressure and desire to be like everyone else will disappear.

It is necessary first and foremost to take care of yourself, whatever that may look like. Being able to manage taking care of yourself is extremely hard work. It is a lot of hard work to rest and recover. I know from experience. I've always tried to get back to normal too early, and that only lengthens my recovery time.

The next thing we can do is to learn anything and everything about our bodies in extremely good detail. Then we can learn how to manage them. Understanding our bodies can allow us to adjust and create new routines, allowing us to prevent additional issues from arising.

When your energy is limitless. there is no second thought when planning your day. Being limited with the energy we have, the things that we do have to be thought out, so we don't over-work our bodies or face the consequences that come with it. Getting ready for the day takes a lot of energy and even more so on days when you need to leave the house for any reason. If we don't take things slowly, we can become exhausted, or end up in bed for the next couple of days. I always have much more energy when I space out whatever needs to be done and that which would take more energy than I'd spend on a normal day.

Even now, I limit my doctor's appointments to once a week, sometimes scheduling as many as five a month. Treatments do help with managing fatigue and some days you will feel better than others. You can always change pace at any time. Even still, I consider factors that affect my energy. I sometimes assess daily and do what I can for that day. As things improve, then it is much easier to manage your energy.

## MANAGING YOUR PAIN WILL INCREASE YOUR ENERGY

There are many different procedures, medications, and treatments that can help to improve your pain and in turn improve your energy levels. With so many different treatments out there, make sure that you research and consider your options. You should also make sure to weigh the pros and cons of anything that can affect you negatively, especially because our bodies are so different and can be fickle.

As a general rule, I go by "everything is worth a try. You won't know what can help and what doesn't unless you try it first!" If there is something that doesn't work or causes a reaction, then it's always good to revisit those things in a couple of years. Our bodies are going to be very different and can change greatly in that time. You never know if your body will react differently later, as your hormones will be different, and your needs will also change. There are several medications that I can take now that I couldn't before. There are also foods that I cannot eat that I could before. It is a good idea to reassess your situation every few years, especially as your pain may increase.

Being able to manage my fatigue and pain better allows me to one, get out of bed, and two, to feel much more like myself which allows me to have room for adding happiness back into my life. I have had friends and doctors tell me that I am a completely different person depending on my fatigue and pain levels. I have a personality that does take quite a bit of energy and a brain that has dealt with sepsis and other infections. Healthy people have a hard time understanding how we really feel when we say we are fatigued and exhausted. I've been told a lot of times how people have a really long day and spend many hours doing physical work, like packing and moving, and as a result feel tired and sore. Before I became symptomatic, I had plenty of days like these, working open to close waiting tables with a short staff and no host to help. I had days where I walked circles around the 10,000-step goal so to speak, so believe me, I know what it's like to be tired and exhausted. I didn't understand the difference between being tired and exhausted and having *actual* fatigue until I became symptomatic. Most people cannot grasp the concept and reality of how fatigue really feels. As a healthy person, I never experienced the type of fatigue that comes with chronic illness.

## SPOON THEORY

Spoon theory is a more tangible way for us to better visualize and explain why we are so exhausted all the time. A lady who had chronic illness for many years came up with a way to try to better explain the difference between being exhausted and having fatigue when her friend could not understand the concept of how she never had any energy. The theory is that a person with disorders and chronic illnesses has a finite amount of energy. Think of your energy being measured as having an infinite number of spoons. You never run out. For those of us with true fatigue, we are only given twelve spoons maximum every day to do everything we need to, from taking care of ourselves, fulfilling home responsibilities, going to doctor's appointments, or going grocery shopping. With a chronic illness, you also are unable to determine how you will feel in the future, so you won't know how you will feel one day until you wake up that day. We must use our spoons differently and on bad days we may have only two or three spoons available.

Spoon theory has examples of how many spoons would be allocated to a list of activities. So, getting out of bed, getting dressed, and taking all our medications might each take one spoon. Bathing, reading, and studying, putting on makeup, and styling our hair each cost two spoons. Driving somewhere takes three spoons and going to the doctor or grocery shopping will take four spoons. With these activities, you use up twelve spoons fast, making it very important for us to plan out everything we do day-by-day. Eventually, we may be able to increase the numbers of spoons that we can spend per day.

As a health coach and fellow person diagnosed with a chronic illness, Tami Stackelhouse said, "I should think of our energy being like a phone with a bad battery. Healthy batteries will charge 100% and last throughout the day, but chronically ill batteries will say charged to 100% but the battery runs out by mid-morning."

## YOU NEED TO DECIDE TO SUCCEED

Take time to decide who you want to be now that life has changed. Most likely your job or desired career is no longer something you can do; it's okay to focus on a new you and new future. For me, getting my doctorate degree in psychology wasn't something I could attain. The longer I held on to that dream, the longer it took for me to change my perspective and ultimately and purposefully choose what came next. Since being diagnosed, my desire to help others has only strengthened. I came up with a way to do just that by going a different route yet still making my goals achievable. You'll be surprised at what you'll come up with as you hold to your desire, decide how to still make that happen, and determine the best way to get there. Don't exclude new and different careers or any goals but realize change can give you new directions or destinations. Our happiness is not contingent on meeting our old goals. I found it much more rewarding to keep to my old desires in a new way, rather than focusing on the exact goal. Coming to peace with your new life is how you'll allow yourself the possibility of future happiness and enjoyment in life.

We are all survivors of our EDS but just surviving isn't living. Surviving each surgery, coma, and near-death experience has kept me alive. There's so much more to living than just getting through the day. Just know that there will be things that we had planned for our future that we may not be able to achieve anymore. So many people feel like this makes us failures or unsuccessful, but it doesn't have to be that way. We can choose to feel sorry for ourselves and merely survive. We can also choose to redefine success in our lives to fit that which we can still accomplish, despite our EDS.

# 5

## LIVING AN EDS LIFE

Hearing about an event and actually being there to experience it yourself are two very different things. It's similar to reading a book and watching the movie adaptation. Our stories need to be told, by us, so others can properly understand our lives. What others need to know isn't just that our lives are difficult but that our lives are extremely traumatic. They are every bit as traumatic as when you live in an abusive house, and we don't get a vacation.

In April of 2015, Dr. Forest Tenant primarily treated cancer patients as a specialist in untraceable pain. He said, "At this point in time, I put EDS in the category of being in the top three or four most severe pain problems. A lot of people, for example, think that cancer pain is the worst of pain but let me assure you that many EDS patients have pain far beyond any cancer patient I've ever seen. It's one of the pain problems that is severe and has been very troublesome. Many physicians are afraid of the disease and of the kind of pain that EDS patients have."

There is no escape from it and there will never be an end.

## HIGH PAIN TOLERANCE

Having the condition since birth makes it hard to discern pain levels sometimes, because for all we knew, everyone experienced pain like we did. Always having pain gives us a very high tolerance for pain. When I was thirty-seven, in the summer of 2018, I sprained my ankle, at least that's what I thought. I iced it, kept it elevated, and only walked when I couldn't avoid it. This sprain happened at the worst time of my life. Looking back though, I know it had a lot to do with malnutrition. It had been five years since I was able to eat without pain and I had yet to find someone who believed me.

Two weeks later, I went to an urgent care for an x-ray. I had an avulsion fracture, which is when a tendon or ligament is pulled so hard that it rips off a big chunk of bone out of your tibia or fibula. After this, I was put in a boot that cut me off from the life I was building after having been bedridden for three years prior, and spending one and a half in therapy. The boot was on my right foot and that's the one I use to drive. Believe me, I tried to drive, and the results where both disappointing and hilarious! I was referred to orthopedics to make sure that my leg healed correctly. It was weeks later when I had my first appointment and was told that these types of fractures take at least six months to heal. It really sucks that I didn't even know of EDS at that point. With so many surgeries with orthopedics, someone along the line should have picked up on it.

This fracture effectively quit my part time job delivering packages. Finally, six months later, there was no change, which was not a surprise. The doctor did what he should have done six months prior and ordered an MRI of my ankle. That took a week before there were available appointments, and then several more weeks to see the orthopedic doctor. When I heard the results, I was so irate! I was in a boot for over a year – stuck again at home alone. On the outside of my ankle almost an entire circle of tendons had been ripped apart; there was no chance of healing without surgery to re-attach everything. By this time, it was February 2019. I had just gotten my first feeding tube, which looked like my nose broke with all the blood. So, I was trying to navigate a scooter around the house and had to go up and down the stairs on my ass. I wasn't used to being attached to a pole and I'd run my scooter over it all the time.

## TILT TABLE TEST

With POTS, if there is not enough blood flow to the brain when someone stands up, you can easily pass out. Additionally, it's well known that an emotional tragedy could make your POTS worse. My husband's grandmother had just passed away, and I was so sad that I didn't get a last little bit of time with her. I also went septic my first time in November 2019, and I was hospitalized for a whole month. COVID would for sure have killed me if I got it. Although we desperately needed distance learning for my son, the high school Logan was in wanted to put him in a raffle for distance learning. I was like, *are you serious?* I almost died and Logan was still finding me passed out all over the house. I told the school to drop him out of school and he graduated six months early. Not only was it terrifying to find me all over the house unconscious, but he was also really scared of going to school because he was worried that he'd get COVID, bring it home, and kill me.

Another interesting aspect of my case is that when I passed out, my heart didn't go into tachycardia like it was supposed to, and I never reacted to pain. Think about it – even people in a deep coma still respond to pain. Every time I woke up, I could never feel my body, just my head. It was horrifying that all this was going on, and still we didn't have any testing done, so all we knew was that I'd pass out anywhere from a few minutes to a few hours. I was able to get a tilt table test (TTT), the only test to measure your POTS.

FULL-TIME JOB

One of the things that makes EDS so hard to live with is that everyone has their own experiences. All our bodies will fail, we can expect that. What we can't expect is how our bodies will fail. Too many factors come into play, not limited to genetic factors. There is an environmental impact as well, as we are all exposed to the minerals and chemicals of our area. The way we were raised is also a consideration, as those of us who grew up in a city have very different experiences and demands on their bodies compared to someone who grew up on a farm. We also all have our own interests, habits, and activity levels; for example, some may get involved in sports, camping, hiking, or swimming.

There is no way to predict how, when, and in

what order parts of our bodies will fall apart. One thing we all have in common is how many doctors we see and how much time we spend in doctor's offices. The amount of time it takes to manage this disorder is equivalent to that of a full-time job – except we don't get nights and weekends off. There are no vacations, no sick days, no personal time, and we don't get any breaks, and the worst thing is, we will *never* retire. The EDS job requires a large amount of secretarial work, like making phone calls for appointments, to insurance agents, filling prescriptions, and a mountain of other tasks.

Most of these appointments require us to wait several months to get in. It is also extremely hard, *physically,* to get to these appointments. Most of the time, we don't know how we feel until the day of and we've all had times when we've had to cancel the appointments the day of and wait several months for the rescheduled time.

Additionally, we all have to find the right treatments that work for us, since it's not one thing that makes us feel better, it's fifty or more things that make us 1 or 2 percent better. Managing our condition most likely takes years to figure out what works best for each of us. We see our doctors so much more than the average person, cutting into time spent with our families and friends. Half the time, when I'm admitted to my home hospital, it is really like a family reunion each time when I see familiar hospital staff.

## PUTTING THE PUZZLE OF YOUR BODY TOGETHER

We all start out with only a few pieces of the puzzle. At first, the more pieces we collect, the more the picture doesn't make any sense. There are so many unknowns, an increase in how many doctors we see, and an unending list of tests we have to take. Each doctor has their own preferred tests for patients. Sometimes multiple doctors want the same tests, so there are times we actually end up taking the same test multiple times. You end up really just feeling like one big science experiment. There are tests to find out what's wrong, tests on medication types, tests on different amounts and combinations, and tests on all kinds of treatments. It's awful when you know that what will help you is out there, but you can't do or try too much. We can easily spend a couple of days in bed if we overdo it. There is a balance we all need to find for ourselves and anytime we get sick or have new symptoms, it starts all over.

## TRACKING, LISTS, AND ALARMS

The most helpful tip I've learned is to track everything. This is how you can learn the most about how your body works. Tracking will help you figure out how everything affects you. It's all interconnected; what you eat can trigger increased pain and other mass cell activation syndrome (MCAS) reactions that didn't seem possible. I've eaten things that have gotten me full on drunk, slurring my words and not walking straight. You wouldn't believe I was 100 percent sober. After eating something that causes this type of reaction, I officially end my night. I'm not able to finish the meal and spend time with my family. I need to be helped up the stairs and end up being put in bed to sleep it off. Knowing your body's triggers allows you to avoid negative reactions and gain back some semblance of normality. The ability to sit down at the table and enjoy a family dinner seems silly, but those are the everyday things that can be taken from us, and tracking makes it even more possible to continue being a part of it.

You'll also live with lists. You'll have a list of things you react to, a list of symptoms they cause, a list on how to avoid the pitfalls, and a list of what to do when you experience symptoms flaring up. You'll have doctors' lists, a long list of medications, lists of what answers you are still looking for, and an ever-growing to-do list. Keeping a list of current medications can be useful for a babysitter, neighbor, family member, and for your partner, especially when you go to the hospital unconscious.

Then there are the alarms that help you remember things, especially for times of heavy brain fog or for, in my case, my sepsis brain. I need alarms two days prior to a doctor's appointment so I remember to rest and prepare for the outing. Everything gets more than a one-time alarm. I set as many as needed, depending on how well my memory is at that time. I used to need alarms several times a day to take my medications on time, and I'd set an alarm every time I started on a project or chores, like laundry and returning a phone call.

## MAKING MEDICATION EASY

Feeling our best takes a lot of different medications. I'm currently taking around twenty different medications. Each medication has to be taken at certain times and missing even one dose can create a list of problems. Skipping one day of depression meds causes my mood to take a plunge and puts me into a very deep depression for three to four days. Our bodies are incredibly dependent on our medication and treatments, and if we don't make sure to keep up on them, it can take us days, sometimes weeks, before we are able to return to the condition we were in.

Other medications have more severe symptoms. For example, neurological meds can leave us with pain, ticks, and painful muscle spasms. Pain can be the hardest to get under control. Missing medications is much more serious than people know. To keep track of the medications I take, I bought four one-week pill boxes that have compartments for four times a day. Every four weeks, I'll fill these medications boxes. Filling them may take two to three hours, but then I have my doses ready for a month. Since I started this, I hardly ever miss a dose of anything.

RESCUE BAG

Another extremely helpful thing I do is use what I call a rescue bag. My favorite rescue bag is a medium-sized clear plastic makeup bag, although you can use whatever works for you.

I have had someone steal a large amount of my pain medication. Because of my luck, I got called in for a pill count the next day, which is when I found out I was short about thirty-five pills. As a result, I was dropped from the clinic, even though it wasn't my fault.

I dropped the thief from my life, had to find a different pain doctor, and most importantly, I needed to figure out a different way of keeping meds on me when I left the house without taking the whole bottle with me and needed something big enough that I wouldn't lose it. Plus, I hated having to figure out what meds I might need with me every time I left the house.

I came up with the idea for my rescue bag having anything I would need to rescue me from any symptoms while out of the house. After my pills were stolen, the only bottle I don't put inside the rescue bag with all of the pills are my narcotic medications; instead, I only take a couple of these with me in the bag. In my rescue bag, I have three medications for nausea, as well as meds for anxiety, Valtrex, and blood pressure. I also have potassium, salt, and glucose tablets. In a small pill holder, I keep a few of all the basics, like Tylenol, ibuprofen, aspirin, Sudafed, Mucinex, and I absolutely don't go anywhere without Benadryl. If you haven't been symptomatic for long, you may not be on very many medications. I've been symptomatic for twenty-three years and I have a large box full of medications. Especially when you're given a three-month supply for most of your prescriptions, it's a huge ordeal to pull them out and hunt down a medication in a hurry. It's been extremely convenient knowing exactly where to find a medication when I need it as quickly as possible.

## NUTRITION AND MEDICAL EQUIPMENT

With EDS, there are several types of treatments that require a lot of medical equipment, including enteral equipment. Two of the most common needs are a peripherally inserted central catheter (PICC) line or port for easy IV access and for IV saline treatment to prevent constantly passing out; some of us require this daily. Another need for a PICC or port is for daily infusions of total parenteral nutrition (TPN), which is nutrition by IV. We are drowning in alcohol wipes, Tegaderm, weekly dressing change kits, flushes, heparin, and so much more. At my house, one of my large closets and dressers are full of supplies.

The second need that comes with copious amounts of medical equipment is our feeding tubes; 90-95 percent of those with EDS do better having one, whether the feeding tubes empty in your stomach or small intestines. A very high percentage of EDS patients have the inability to digest and absorb foods and experience pain as a result. The upkeep needed for enteral equipment just eats away into our time and energy. It's a vicious cycle, but our lives truly depend on the medical equipment we need.

I get four very large boxes a month, and this contains all the nutrition I need. My body had also quit growing my hair and nails. It caused me to go grey in my early thirties. Additionally, my body was so malnourished that twice I fractured bones when I fell. People with EDS could have the healthiest meal plan, and stick to it 100 percent, and still be extremely malnourished because our bodies simply can't break down the food, and with a common EDS comorbidity, mast cell activation syndrome (MCAS), we react to so many things we eat, and it only makes us feel worse. Having a tube or TPN takes away the need for your body to break down food and we survive off formula.

There are health insurances that won't cover the cost of formula – they will only pay for all the equipment. At a time, it was costing us an additional $640.00 for a month's worth of formula. For those of us who are able to eat orally, it takes a lot of work. There is no last-minute stop at a fast-food restaurant to grab a quick meal. Having MCAS, there are way too many foods that will cause a reaction and it's not just an anaphylactic reaction. We can have a hard time consuming food and it gets expensive when there are only so many options that don't cause pain.

The people who are the most reactive to foods will rarely eat something that they have not made themselves. Having bad reactions isn't the same as not being hungry. Even though there are times when our dysautonomia forgets to signal our brain that we need food, it doesn't change the fact that we all still need nutrition. Your body may not like it because essentially these foods are poisonous to us, but it won't take away our desire for our favorite foods.

I have struggled more with this mentally and emotionally. When I was a teen, I was anorexic. Eating disorders are just as hard to overcome as alcoholism – there are even rehab facilities for it. I have struggled with this again since 2013, when it became so painful after eating that I'd be in the ER for pain control. No amount of the pain meds I have at home can safely help me have the pain back under a manageable threshold. I knew I needed nutrition and that the only way I could do that was through a tube and it's felt like a forced form of anorexia. In a way, I though it wouldn't be so hard. There are times when it's worth eating a favorite food and I am prepared to deal with whatever outcome comes. When I learned of a new eating disorder, avoidant / restrictive food intake disorder (AFRID), it helped put things into prospective for me. AFRID includes avoiding food to avoid pain and having little interest in food. This newly identified disorder just confirms that I have had an eating disorder for the last eleven years. There are five types of AFRID, which are all worth reading about, especially if you have a hard time with nutrition.

## HUMOR AS A COPING MECHANISM

Humor is how we get through every day, because if we don't laugh, we'd constantly be crying. I still have weepy days but overall I have the ability to just laugh at the mountain of issues I face. When I got my first feeding tube, I named the pole – that was to be connected to me for a large portion of the day – Stanley. Stanley just stands there and holds everything I need. When I was given a PICC line, I ended up with a second pole for IV saline. My husband told me it was Stanley Two. "Stanley Two?" I asked. "That's the worst thing to name it, absolutely not." We compromised and named him Stanley Tucci instead! I was also given a walker to help me, especially during physical therapy, and I decided his name was Christopher Walken.

## SLEEP AND INSOMNIA

Just as important as nutrition is sleep. It would be heaven if I could sleep every night. Insomnia is even worse than I could have imagined before I experienced it for myself. I've been up all night, so I thought I knew how to deal with a little sleepiness. But insomnia isn't just not being able to sleep one night. Insomnia is relentless. You could suffer from insomnia for one night every week, or several nights in a row. After a while, it becomes a regular part of your life.

I experienced the worst repetitive insomnia that left me more or less brain dead. I was awake for 100 hours, with random naps of an hour at most scattered in between. That's four days awake with maybe a total of three hours of naps. When I was finally able to sleep, I slept for twenty-four to thirty-six hours straight through. There was no way to wake me up.

I also experienced this while I was inpatient, and my body would just shut down. I couldn't physically keep my eyes open, no matter how hard I tried, and no matter how much the doctor tried. I didn't even respond to pain – and even coma patients will still respond to pain. Apparently, I went through as many tests as doctors could come up with – blood, urine, X-rays, CT scans, multiple MRIs, endoscopy, and ultrasounds. When I woke up eight hours later, I was on my way to the ICU. Those poor guys were scared silly; they knew what had been going on and here was this supposed-to-be vegetable suddenly sitting up and asking, "Where are we going?"

I'll never forget scaring them. So, when my body crashes, it crashes hard. This cycle of 100 hours awake followed by twenty-four to thirty-six hours of sleep and a few short days of normal wake / sleep hours – I'd start the cycle over starting with no sleep for days. The first day or two, I was mentally no different than any other normal day. Around the forty-five-hour mark, my cognitive function essentially became non-existent. I couldn't understand simple concepts. I couldn't think straight. I was easily confused, and I wouldn't know where we were if we left the house. This cycle happened for almost six months.

Along with insomnia, sleep disorders are common. I have one, but I'm just being tested to determine *which* one. The symptoms I have may not sound scary, but while I'm having these episodes, it's absolutely terrifying. I try to keep a consistent bedtime, as we all get our best healing while we are sleeping. I never know when I'll have these episodes. So far, I haven't been able to connect a trigger to it, but then again, there may not be one. Sometimes, after I fall asleep for several hours, I wake up suddenly like a jolt. Only when I wake up, I find myself standing in the bathroom in front of the sink – with the light on. Being startled awake in the first place is one thing, but finding myself in front of a mirror alone is quite terrifying. It felt like someone was right there in front of me. I've finally been able to figure out to grab for the sink immediately, because if I don't, I fall back flat on my ass.

My tailbone had been badly bruised recently when I passed out without a warning while I was going down the stairs. My behind was black for months. So, if I don't grab onto the sink for stability, I'll fall backwards and land on my already injured tailbone. When I either caught on or fall, everything goes black. I don't even know what happens, but my best guess is that I went to sleep, I just don't recall doing that, or even leaving the bathroom. In what seemed to be ten minutes later, I'd jolt awake again and I'd still be in the bathroom, standing in front of the sink and mirror again. I'm not sure how long it lasts, and I can't keep track of how many times I do this every time it happens. My best guess is five to ten times.

Eventually, I'd wake up and be in my bed and it would be morning, making it difficult to judge how much sleep I actually got. Then again, it may not matter, since I wake up exhausted. All of this makes it impossible to really get to a point where I have good sleeping habits and sleep regularly for a long time.

## SEX

When I say sleep disorders take over everything in your entire life, that includes sex. You'll have to get a bit creative. Pain and energy make it hard to be spontaneous. Not knowing if your body will do what you want it to, neurologically speaking, at times you won't have the best control over your fine motor skills. Our symptoms change randomly, and you might experience muscle spasms, your body overheating, or your body being so sensitive that it's painful to be touched. This makes it almost impossible to set aside certain times for sex. Then POTs can cause issues with getting dizzy and possibly passing out. You're always hoping that you'll be finished having sex before doing so. I would not have sex all over the house when our kids lived at home. Since I didn't know how long I would be passed out for, I didn't want my kids coming home and seeing me completely naked and unconscious on the floor or the couch. It's very important for you to communicate this with your partner, because you must be on the same page about where you are physical and beware of your limitations in those moments.

## PASS / FAIL VS. GRADED HEALTH

I have yet to meet someone with EDS who has not been bedridden for any amount of time. I was bedridden from 2015 to 2018 when I was down with a few autoimmune disorders, intense pain, and lethargy. Sometime in 2017, I went to a functional medicine doctor. I'd heard great things about his reputation, and he was a highly-regarded autoimmune doctor. He told me that in the medical field, everything is thought of in terms as a pass or fail. You either fit in the tiniest window qualifying for the diagnosis, or you don't quite meet the criteria, in which case, you absolutely don't have whatever it is you were being tested for. I learned that with regards to our health, we should look at it in terms of grading our illness. I had been tested for lupus and Cushing's disorder, but the results came back that I didn't quite fit inside their parameters; that meant I absolutely didn't have either, but doctors couldn't tell me why my body was retaining so much fluid and they couldn't explain to me why I felt so awful. The doctor told me I was fine according to this test, and he had no further answer for me and most likely couldn't treat me for anything. I was left not knowing where to go next or even what to do next. I was at a loss, confused, and left with only more questions.

When I'd made that first appointment, he was so booked that I had to wait for nine months, and I'd do it all over again. He taught me about how he views the medical field on a graded scale, and told me I was one of the sickest patients he'd ever seen, even though the other doctors claimed I was just fine. One of the first things we went over was my stress and hormone levels. He handed me a paper that was obviously for a graph but there weren't any lines on it. It showed where my levels were. How could he say that he had results, when the graph was obviously blank?

Being used to the pass / fail side of medicine, I tried to prepare myself to be brushed right back under the rug. To my surprise, he explained to me that my body had been in a state of fight or flight and my cortisol levels were so high (which is Cushing's disorder) for so long that my body had quit making cortisol (which is Addison's disease) and all the other required chemicals and hormones my body should be making. All the results were right there; across the board, everything was holding steady at 0 percent, non-existent. I was hormonally flat-lined. I was also just barely missing crossing the line to officially giving me the diagnosis of lupus.

"Here's why we should use a graded scale." The doctor said. "Since you don't have lupus, you should be 'just fine' to other doctors, but clearly, looking at you, you are definitely not fine! People who feel 80 to 90 percent in terms of their health are probably feeling pretty good, and don't need much help medically. For people feeling somewhere in the 50 to 60 percent, they aren't doing so well, and they will deal with symptoms. But, in the pass / fail system, these patients won't be able to get the treatment they need because their lab results are one point off. People aren't incredibly healthy, then one day later, they are so ill they are bedridden. It just doesn't work that way. If it did, then the pass / fail system would make sense."

So, I had one doctor saying, "It's not lupus; you're fine," and another doctor saying I was so close to the windows for a diagnosis that I had options and would benefit greatly from still being treated. After that, I learned how everyone with chronic illness and chronic pain should be treated by graded systems and see a function.

## BEING BEDRIDDEN

From February 2020 until fall of 2023, I was bedridden again. I was in and out of the hospital, with twenty-five of those months hospitalized. During the time between hospital stays, my body was still recovering. I read somewhere that for every day you are in the hospital, you would need two days recovery, so the rest of the time while I was home, I was still in bed for most of the time because of how sick I was. After all the times when I almost lost my life, I didn't just bounce back from that.

It's known that while you are in the hospital, you will be pretty isolated most of the time. Being bedridden is also very different than being in the hospital. When you're bedridden, you hear and see what everyone else is doing, making you feel like you're missing out on all the fun and activities. At least when you're in the hospital, you are unaware of what is going on at home. One of the worst feelings is being left out. Being home, you kind of feel like a bystander. You're there, but you're not really involved in a lot that goes on.

Since we know at some point in time we will be spending time in bed, it's important that no matter how long we are down, we keep our minds going. You're already at risk of losing memories; it wouldn't be much of a stretch to lose some of your cognitive function. One thing I've learned that helps keep my mood and self-esteem up is keeping my body clean, however you do that. Shower when needed or use bathing wipes, or switch between the two. Maybe you'll have a shower one day and use wipes the next. It doesn't matter what or how you clean your body, but don't let three, four, or more days go by without washing your body.

If you are spending a lot of time in bed, then keeping your sheets clean is a need. If you normally wash your sheets once a month, you'll want to wash them a little more frequently because you are spending more time in bed. I have two sets of sheets for my bed so I can have fresh sheets every week but only need to wash them every two weeks.

Additionally, no matter how you feel, brush your teeth and get dressed. I didn't realize how important it was to get dressed every day until I stopped wearing pajamas every day. You don't need to wear fancy clothing and definitely be sure to dress comfortably. I know, I sound like a mom teaching her children (at least that's how I felt when I was told to do it). It doesn't matter if mornings are so hard that it takes you until noon to get dressed. If you are one who has just been in PJs all the time, just change your clothes every day and you will notice that this basic task will have an effect on your mood.

After you have that mastered, if you are anything like me, then occasionally do something that makes you feel pretty, whether it is putting on make-up, trying a new hair style, buying new clothes, or painting your nails and toes. Staying clean and doing whatever it is you do that makes you feel good about yourself, even if you don't leave the house, will have an effect on your mood. Remember, you're not doing this for anyone else; you do this for yourself. Right now, there's no way I have it in me to wear make-up everyday – it's just not going to happen, so I do what I can. I really don't know why these things are important, I just know that I feel like a person on the bed instead of only feeling kind of grungy. Most importantly, stay clean, and then occasionally feel pretty.

## STAYING BUSY IN BED

While recovering from sepsis, my brain had a very hard time making sense of anything. I knew my brain wasn't damaged or dead, but I could feel my mind struggling to remember things and how difficult it was for a while when I couldn't carry on a conversation for longer than five minutes at a time. Your body may not be working, but your brain still needs to be exercised regardless of your physical condition. There are so many things that you can do from your bed. As you start using and exercising your brain, you will come up with ways to accomplish things even from your bed. I did – I wrote this book! Some of the things I enjoy doing for my brain physical exercises are: 1,000 or 2,000-piece puzzles, building with Legos, coloring, doing word searches and crosswords puzzles, cross-stitching, crocheting, and reading. There are so many options for you to choose from that you can do to keep busy while you're in bed.

Learning how to adjust to the new way of having to live your life isn't easy, especially if you have no idea what to do or who to trust. Start with trusting yourself and listening to your body. Once you start paying attention to your body, you will figure it out quickly. Getting nutrition, however you need to, will help keep your MCAS reactions to a minimum so your body can function at its best. It's a combination pain and MCAS reactions to food that lead so many of us to the feeding tube or TPN, both of which take a lot of work and upkeep. It should be a last resort. If you are still able to intake food orally, then I'd try drinking the formula before you investigate a feeding tube. Additionally, sleep when you can; this is the time when your body heals. I'd argue that consistent, regular sleep is one of the biggest challenges we face.

Also, pay attention to your sleeping patterns, as this can help you know if you need testing for a sleeping disorder like narcolepsy. You can help yourself by taking care of you in the way you need to be taken care of – just be prepared that it may take years to figure out how to do that. You'll learn what works for you, one thing at a time, and there will be much trial and error. Let me tell you that all this work is worth it. You will figure it out long before your life is over and everything you do will affect everyone around you – your family, friends, and people you meet throughout your life.

## 6

## EDS LOVES COMPANY

In the previous chapters, I mentioned other disorders I suffer from, along with EDS, like sleep disorders and POTS, to name a few. So, by now, it's no secret that those with EDS also find themselves with other conditions, also known as comorbidities. In this chapter, we'll dive into *why* this occurs and what you can expect with an EDS diagnosis.

EDS is a disorder that hinders the proper production of the protein collagen. Collagen helps keep our bodies together by holding joints tight and holding our bones in place, our teeth in our mouths, and even our eyes in our faces; it just keeps everything tightly together. Collagen is found in everything – in all our organs, teeth, and nails. So, with EDS and producing collagen incorrectly, it weakens the ability for that protein to hold on tight and to be elastic as it needs to be.

Logic follows that with EDS and defective collagen, you will have other issues as well because you are using building blocks that are defective. This is why, when you get a diagnosis of EDS, you will not only get one diagnosis, but you will also get a diagnosis for several things all at once.

## COMORBIDITIES

Those affected by EDS will usually start showing signs and symptoms during life milestones during which there are hormone changes. The two biggest culprits are during puberty and pregnancy, although a lot of patients can have the disorder and not have those symptoms present until they experience a big illness, like an accident. Although I had some symptoms since my first pregnancy, it wasn't until I went septic for the first time at thirty-nine that I had a heavy onset of additional symptoms.

The structural deficiencies of EDS can cause so much damage to our organs and tissues, making it impossible for your EDS body to function like a healthy body would. Essentially, our bodies are like ticking time bombs that will eventually explode. Something will go off, you just don't know which bomb and when. While our fragile bodies have to bear the normal wear and tear of functioning, the structural damage of EDS brings a mountain of comorbidities. EDS is always seen with postural orthostatic tachycardia syndrome (POTS) and mast cell activations syndrome (MCAS) as the two most common comorbidities. This trifecta is seen in all thirteen types of EDS.

Some of the typical comorbidities and diagnosis you may have already gotten include GI issues, neurology, sinusitis, interstitial cystitis, fibro, arthritis, joint pain, dislocation, insomnia, depression, anxiety, inflammation, allergies or sensitivities to foods, metals, and adhesives, to name a few.

While recovering in the hospital, I started to have episodes where I would pass out. I had nurses find me on the floor in the bathroom or next to the bed. My fainting episodes happened in different places until I was instructed to just stay in bed unless I had help. I did not know the link between my symptoms and passing out was POTS and I did not know they were also linked to EDS at the time.

When I went home from the hospital, I was still experiencing episodes of passing out; sometimes, they were instantaneous, while other times I could sense that I would pass out. There are a lot of times when I have felt the episode coming on. With this ten-second warning, the first thing I do is get myself somewhere safe and get down low to the ground. The next thing I will do is try to contact someone to come look for me. In the beginning, I was passing out between eight to ten times a day.

At times, I would pass out and come out of it within five to ten minutes later, but there were other times when it has taken me a lot longer to wake up – the longest being four hours. The first time I passed out, I didn't wake up until two hours later. When I woke up, first my eyes would open, but I still wasn't able to move or speak. I could only move my eyes and blink. I was told when I first opened my eyes that I just had a blank look on my face, and I didn't see or respond to anything going on in front of me.

All of this scared me because I could not feel my body and I was a little confused as to what had even happened. Doctors did a lot of tests at the hospital because I couldn't be woken up. I wasn't even responding to pain techniques that medical caregivers used, which was surprising, because as I've mentioned before, even deeply comatose patients would still respond to the pain. As for me, of course all tests came back normal. I was scheduled for a tilt table test, also known as a TTT. This test is used to simulate one going from laying down to standing quickly, in a controlled environment, hooked up to the appropriate machines. My TTT results showed my blood pressure dropped within the first three minutes of standing up, and I passed out on the table I was strapped to. It then took forty-five minutes for me to open my eyes again. This is when I learned that my heart rate stayed the same for the entirety of the test, when my heart should have gone into tachycardia to speed my heart rate and help recirculate my blood so I could regain consciousness. Since my body doesn't do this, this is why I can be passed out for long periods of time.

## DYSAUTONOMIA

POTS and MCAS are disorders that fall under the umbrella of dysautonomia. Dysautonomia means any disorder of the body's autonomic systems. Your autonomic nervous system takes care of all the things your body does without thinking about it, like temperature control, blood pressure, heart rate, digestion, blinking, feelings of hunger, and your allergy, immune, and stress responses. All of these things are affected in a big way. Symptoms can fluctuate all day, making it more difficult to choose what you wear that day and getting the warmest blanket to stop shivering, only to take it off every few minutes when your body can't regulate your temperature. For those who wear makeup, it's almost impossible if you can't stop blinking. It's not just the big symptoms that are hard to deal with; the small ones can be very frustrating and irritating.

## POTS

One of the most common links to EDS is a condition called POTS. This syndrome has to do with inappropriate blood circulation in the body and most likely, most patients experience the same symptoms collectively. This syndrome is caused by our blood pooling at the bottom of our bodies instead of circulating. Standing up and other quick movements will cause your blood pressure to tank, causing dizziness and syncope. If you do happen to pass out, then your heart goes into tachycardia trying to recirculate the blood. This syndrome will affect your heart rate, blood pressure, and body temperature.

POTS can cause your blood supply to be low. Hydration helps in aiding to increase your blood supply but is still different than your overall blood supply. Our heart rate (HR), blood pressure (BP), and temperature will fluctuate greatly throughout the day. Our HR could quicken and our BP could tank, causing you to pass out. Our bodies are also incapable of controlling our temperature; most of the time, we're usually cold, but when we get too hot, our bodies will stay hot and possibly end in heatstroke. Staying hydrated and wearing compression clothing can help our bodies retain our blood supply. We are all generally on a high sodium diet. There are times when it's impossible to keep our blood supply high enough to keep from passing out – several times a day. The weather and celestial events can cause a flare up of symptoms. If symptoms don't improve, we rely on IV hydration to increase our blood volume. When I was passing out up to ten times a day, I needed IV fluids on a daily basis. The most common way to get these fluids was to have a PICC line placed. Remember my hygiene tips, and if you have one, you must be very careful and keep it sterile because it is easy to get an infection.

## CELESTIAL EVENTS

POTS symptoms will also flare up during celestial events, such as during a solar eclipse, full moon, solar flares, meteor showers, yearly equinoxes, conjunctions, lunar eclipses, and oppositions. Biometric pressure and the atmosphere are inescapable! Just for curiosity, I decided to look at the last few years to see how this affected me. Everything lined up with times that I passed out and injured myself.

During the lunar eclipse in July 2018, I tore all the tendons in my ankle.

During the Venus and Jupiter conjunction in February 2020, I fractured my ankle.

During the Lyrids meteor shower in April 2021, I fractured my T7 and T8 vertebrae.

And during the Perseid meteor shower in August 2021, I had an eviscerated vertebra.

## MCAS

MCAS, as you know, is mast cell activation syndrome. Our bodies are hypothetically allergic to anything. Taste, sight, sound, and touch can cause an MCAS reaction. The reaction from taste is very common. MAST cells are our bodies' allergy response. When we encounter things, our bodies can react allergically, and the response will be 1,000 times bigger than what a normal allergy response would be. Things we commonly react to are latex, adhesives, metals, foods, loud sounds, flashing lights, and smells. We are also highly sensitive to having reactions by touch. I personally have had a reaction every time I go down the laundry detergent and cleaning supplies aisles. I haven't gone down one in almost twenty years.

Comorbidities that are linked to MCAS have

symptoms that can be kept under control and at a minimum when you limit your exposure to anything that your body has an adverse reaction to. Two of these conditions that have potentially debilitating symptoms are mucosal shedding and interstitial cystitis. Mucosal shedding, mucosal peeling, or oral epitheliolysis is where the superficial layers of skin can peel off and can be painful. When I had mucosal shedding, the areas were white or even clear, making it very difficult for anyone else to see. I was put on a steroid to help with inflammation and an antibiotic, both of which did very little. What really made the difference was throwing out my toothbrush and changing the brand of my toothpaste and mouthwash. Replacing everything that I used for my oral care allowed my mast cells to calm and all the shedding cleared up completely.

Interstitial cystitis is when you have all the symptoms of a urinary tract infection, without having an infection. These symptoms can range from mild to severe. There is cramping, urgency, and burning while urinating, and the urination can appear cloudy and have a strong odor. I used to have severe symptoms and was put on a daily antibiotic as well as a numbing medication to help with the pain while urinating. On days when I forgot to take these medications on time, I'd start experiencing severe symptoms within a few hours. I'd then have symptoms for the rest of that day, and I'd continue to experience symptoms until I'd taken the medication for two consecutive doses. After taking the medications preventatively for a year and a half, I was able to discontinue taking the medications after I altered my diet and quit eating foods that my body was sensitive to. I'd stopped taking the antibiotics and been symptom free for several years, though there are times when I found a food that I have developed a sensitivity to by having substantial symptoms the next time I pee.

Endometriosis is a well-known condition that can have severe symptoms especially when it spreads. It can grow on other organs, only increasing symptoms, including not being able to have children. Endometriosis is also made of mast cells. Knowing what specific things your body reacts to can help you to avoid those things. There are tests out there that you can buy to test your blood, and they will send back a breakdown of foods that your sensitive to. You can also research nightshades and research what the products you are using are made of. For example, potatoes are a nightshade vegetable. The high starch content in potatoes is something that can be added to products like lotions, shampoo, and other health, body, and beauty products. Another thing that isn't obvious are products that say "natural flavors," because they don't reveal the list of what those are and you may have to call the manufacturer to get those details.

Reactions range from rashes, to asthma attacks, to anaphylaxis, to nausea, to inflammation, or itchiness. We also have odd reactions to foods. We can become sick and vomit, get a rash, feel sluggish, experience increased pain, or suffer from severe brain fog, similar to Alzheimer's and dementia. We can also experience stomach problems, IBS, and sometimes diarrhea. Symptoms can onset very quickly. Because of our reactions to foods, whether it be pain when eating or severe IBS and anaphylaxis, 90 to 95 percent of those with EDS do best with a feeding tube or TPN.

Mast cell reactions are not always anaphylactic in nature. Severe mast cell symptoms can cause a brainstorm, which is an electrical storm in your brain and can change the chemistry of your brain like epilepsy. You may also experience seizures, color auras, hallucinations, unfound stroke symptoms, and tremors that are intermittent or have unfound causes. I have said for a long time that when I have severe brain fog, I feel like it's much like dementia and Alzheimer's. I learned that this is actually being in a dissociative fugue state. This is a reversible amnesia.

It is common for someone to wander and travel. You can also exhibit emotional and physical manifestations that are new, with new behaviors and mannerisms. I've experienced this twice after a near death experience caused by severe antibacterial resistant infections that put me into a coma. The only other thing I experience during this state is the loss of muscle control and the muscles I can't control are the ones that help control my bladder, which is unfortunate.

## CCI AND AAI

There are a few comorbidities that are more serious and much more difficult to live with. Craniocervical instability (CCI) is a result of the tendons and ligaments in your neck being so lax and weak that they can't support the weight of your head. Your head crushes your spinal cord and all those nerves. Some have a little instability and can find relief of symptoms by wearing either a soft or hard neck brace. In severe instability, there is a surgery that can be done to help lift and stabilize the head. This surgery is very complex; in fact, there are only five surgeons in the world that will currently do this operation. The two I know of are in New York and Spain. Also, to ensure that no nerves are damaged, leaving you with permanent symptoms, it is best that you stay conscious while having the operation. The incision is long, and minimum recovery is one year. Atlantoaxiail instability (AAI) is the same condition as CCI, except the vertebrae that are instable are lower down your spine. Symptoms range from mild to severe vertigo, memory loss, chronic headaches, neck pain, numbness and problems with vision, to name just a few.

## TETHERED CORD

With CCI and AAI, most cases will also have a tethered cord and will have to have another operation to correct it. Tethered cord syndrome (TCS) affects the lower vertebrae of the spine. This comorbidity causes your spinal cord to stretch and can get attached to surrounding tissues in a way that damages or pinches your nerves. The symptoms are numbness, loss of sensation, weakness, and sometimes, your bladder control is affected.

## CHIARI MALFORMATION

Another more severe comorbidity is Chiari malformation (CM). This is caused by the cerebellar tonsils in the back of your head that herniate from the brain into your spinal canal. This can block the spinal cord and put pressure on the brainstem and cerebellum. The cerebrospinal fluid won't flow correctly and can cause a leak. There are risks of paralysis and stoke during surgery.

## AUTOIMMUNE AND RAYNAUD'S

Another type of comorbidities are autoimmune disorders. This group of disorders are autoimmune because they are caused by your own body attacking its own cells. Anyone with an autoimmune disorder is automatically "allergic" or has a sensitivity to gluten and dairy products. One of the most common autoimmune disorders seen with EDS is Raynaud's. Raynaud's is a circulation disorder where your body has a hard time keeping your blood flow at a high enough level. The most noticeable and common symptoms of Raynaud's are fingers, toes, hands, and feet turning white, blue, and sometimes purple when they are cold. Your extremities don't get sufficient blood flow to keep your pinkish coloring. Another common symptom is having your blood pool. Your body can become very red and splotchy when blood pools in one area. There are things you can do to help encourage circulation like massaging those areas and taking precautions to prevent overheating.

## FASCIA

Our connective tissue is called fascia and it's throughout our whole body. Fascia is made with several layers of a web-like tissue that's made of a protein called collagen. The fascia helps keep our bodies together by surrounding every muscle, organ, ligament, bone, joints, nerves, and every cell. It helps our muscles to contract, stretch, and hold all our organs intact. Healthy fascia is smooth, silky, and slippery, allowing all your organs and muscles to glide by each other during movement. Like anything else in the body, your fascia can become unhealthy. It becomes thick, clumpy, tight, and sticky, all of which can cause a considerable amount of pain. Fascia can be torn, tight, restrictive, and get adhesions. The damage is caused by overextending your muscles or limbs, dehydration, poor sleep, low activity, surgery, injury, and stress.

The idea of people with EDS having sticky fascia is a theory. With EDS, it's certain that our fascia isn't as strong as it should be, which is evidenced by how long it takes for our fascia to heal. Christie and I are of the opinion that people with EDS, or any connective tissue disorder, have sticky fascia. We think it is sticky, and not as silky and smooth as it should be to help your organs and insides slide easily around each other. When our fascia has been injured, it can become extremely tight and can pull organs and bones inward. Without the proper treatment on your fascia, the tight hold on your body cannot be released. There are several different modalities to treat fascia, but be cautious in the ones you try. Our bodies are all so different and it is possible that one modality may set you back a little yet work for someone else with EDS. Unfortunately, with our condition, everything is done by trial and error to see what methods and modalities will work and which ones don't. I recommend doing a lot of research on the different modalities, taking it slow with the ones you feel might help you the best, and if you're unsure, learning how to use and relying on muscle testing your body. Muscle testing is something I learned from a chiropractor and is a valid form of learning to be in tune with your body and knowing what is best for you.

Some ways to help treat your fascia are to use heat therapy, which you can do at home, but steam and infrared saunas work as well. You can also do the opposite and do cryotherapy or use ice packs. Stretching and mobility exercises like low-impact aerobics and yoga can help relax and loosen fascia. Knots and adhesions can be treated by using pressure and trigger point techniques. Massage, foam rollers, acupuncture, and also handheld tools can be used to treat the fascia yourself.

## OFF-GASSING

Think of any brand-new item – clothing, flooring, plastics, literally anything. Some items have a stronger "new" smell, and this is because it is emitting toxins and chemicals into the air, which is called off-gassing. Foods and other natural things are also off-gassing but there is a difference between safe and toxic off-gas from the item. With MCAS, we are more sensitive to what is being emitted into the air. Where all-natural things don't cause any problems for most, those of us with MCAS can have a reaction to both toxic and natural off-gas. We all know that we can easily have a reaction to even the most unlikely, low emitting things. The smells that are stronger are emitting the most, like the inside of a new car, or a new carpet and the adhesive used to keep it down.

I had been given a new hoodie that I wore for a few hours one night and again a few hours in the morning. In this time, my face and both of my hands got a bright red rash that I couldn't stop scratching, even hours after I took it off. I washed the sweater and with a little Benadryl to clear my rash, I could wear the sweater and it never happened again. I learned the hard way that you just wash everything before you use it for the first time. All other items just need to be set out for at least twenty-four to forty-eight hours and longer for the things that have a strong smell. If it's doable, you can set your stuff out to be in the sunlight if you want. Architecture constructing, home decorating, and other similar careers have known about this for a long time because these volatile organic compounds can have a negative effect on anyone, not just those with low immune systems, babies, and us who are genetically challenged. These chemicals and toxins have the potential to cause headaches, asthma attacks, neurological symptoms, and respiratory illnesses. Some more severe effects of long-term exposure can be to your lungs, your nervous system, your cellular makeup, or cause disrupted hormones. These effects have been proven and if you are left wanting more information, there is a database and organization driven to help consumers understand product composition, called Mindful MATERIALS.

GASTROINTESTINAL

Along with EDS is a long list of gastrointestinal comorbidities, including gastroparesis, inability to digest or absorb nutrients, severe IBS, and diverticulosis.. There are a lot of symptoms that overlap; for example, most of these illnesses cause insomnia. The worst thing I've learned about is getting C. diff which happens when you take so many antibiotics that you kill some of the good bacteria that you need in your intestine. It's one of the hardest things to recover from and for those who can't seem to get over it, there's always a fecal transplant.

With EDS, the integrity of our body is compromised at the cellular level. This "defect" causes us all to have some of the same comorbidities. The ones that are almost always present are dysautonomia, POTS, MCAS, GI problems, and sleep disorders. However, there are others, like Marfan's, Eagle's Syndrome, thoracic outlet syndrome, and skeletal structural issues. Comorbidities that are commonly seen together with EDS are endometriosis, diverticulosis, ADHD, autism, small intestinal bacterial overgrowth (SIBO), and feeling of pancreatitis and pain soon after eating or severe pain in your abdomen and GI tract.

You will learn a great deal from others with EDS. I have learned something that I didn't know before with every person I've talked to. Research and learn anything and everything that you can – it may help you or someone you know in the future.

# FINDING THE RIGHT DOCTOR

## THE DOCTOR YOU NEED

There are three types of doctors you will need on an "as needed" basis:

- Type 1: For illnesses like ear infections, colds, etc.
- Type 2: For surgeries or procedures (e.g. I've many colonoscopies; me and both my children have had their tonsils taken out)
- Type 3: On an emergency or urgent basis (e.g. You get appendicitis, pancreatitis, and the doctors you have on an in-patient capacity).

Sadly, the majority of EDS patents would rather deal with a medical emergency at home on their own due to how we are perceived. If it's not life or death, if we go to an ER, we may be there for several hours, and then who knows if you'll get any treatment. All I needed was IV fluids and I was denied the "salt water." In my case, I've spent more than half the last three to four years as an inpatient. Unfortunately, we can't protect ourselves if the doctors don't believe us. I can't share these stories, just yet, but they are appalling and sometimes it's abuse. I have a hospital that I'd rather die than go to. The treatment there was so bad, I signed an against medical advice (AMA) and left.

RED FLAGS

While you consider whether or not a doctor is the right fit for you, here are some things you will want to look for to help you decide.

I recently interviewed for another GI, since my great one retired, and I knew within the first few minutes that I would never see her again. She had known me all of five to ten minutes, and her tone was very clipped. She was close to outright being aggressive. That is red flag #1.

I know she wasn't listening to me because she

was dismissive to several of the problems I've had and how long it was before I got any real help, and she gave me ultimatums. Red flag #2.

Next, while she was going through my medications, she told me to work with her, I had to stop one of my medications or she wouldn't treat me. Having no prior treatment from her, she should not be changing my medications without consulting the doctor who prescribed it. Red flag #3.

She made a negative comment about the doctor I'd been working with for months. Doctors shouldn't be putting other doctors down and questioning the other doctors' motives. In general, you shouldn't have to put another doctor down for them to do their job. Red flag #4.

She continued by insinuating that she was the smartest person, and was not open to listening to what I already learned about my body. Red flag #5.

One of the other ultimatums was to drop the nutritionist I'd been working with for the last five years. Red flag #6.

Red flag #7 was that she wanted me to only consult with her approved doctors. She didn't know me, and I know she didn't understand my body. Making me drop an excellent nutritionist who understood the water weight imbalance would have been detrimental to my health. I shouldn't have to drop a member of my doctor team and all the work I'd done over the past eleven years to get nutrition to start from scratch. Red flag #7.

Her last ultimatum was that I was to work on figuring out in "creative ways" how to get nutrition and to take away my tube feedings, which is the only thing that helps me keep my body from retaining water. Even when I told her that the last time I stopped my tube feed for two days, I had gained 40 lbs., she still recommended "creative ways" to get nutrition. It's not fair to be forced to throw everything out and start over; my body can't take it. Red flag #8.

When I told her that I had EDS and after eleven years of not functioning, it wasn't going to all of a sudden get better because of her treatment, she still wasn't listening to me. She insisted that none of her other EDS patients were on a feeding tube, expecting my body to work like every one's else's does. Red flag #9.

After the appointment, I felt like crap and like I did something to wind up in the principal's office. I was also more confused and torn up about it. Red flag #10.

Finally, she didn't help build a care plan that would work for me. Whatever "care plan" she was going for, it was obviously not the care plan for me. Red flag #11.

Finally, she cut me off several times. Red flag #12.

I do want you to know that with someone "normal," she may not start off aggressive. Honestly, I think it was because I didn't want to give up my feeding tube. This problematic doctor was actually one that one of my other team doctors absolutely loves. In fact, they gave me the referral. This is a great example that, while this doctor may be a really good doctor, she just wasn't the doctor for me.

In these instances, there is nothing you can do to "stand up" to doctors and no way to convince or demand a doctor to do anything. It's awful trying to get help from someone when they don't allow us to help teach them what we know about EDS and our needs. A doctor we want to go to will listen to the expert on your body: *you*!

## TEAM DOCTORS: TYPE 1 AND TYPE 2

Now that you know what to watch for with red flags, it's time to consider what you do want to look for when looking for a team doctor. There are two types of doctors who work with you. The first type is a doctor that knows nothing about EDS and is okay with it, but will help you anyway they can. The second type of doctor will be able to help you manage your different problem areas, like a GI for feeding tubes, or cardiologist, neurologist, or dysautonomia doctor to help manage POTS and other related issues.

## GREEN FLAGS

When you see a possible team doctor, that doctor sets you at ease. You will be able to know really quickly what atmosphere the doctor sets. Green flag #1.

Watch for general respect and a doctor being present and courteous. Green flag #2.

A doctor who listens to what you are saying is green flag #3.

A doctor who allows you to finish talking? Green flag #4.

Going to an appointment, you will surely take your diagnosis paper with you. The doctor who accepts your diagnosis with no questions asked is green flag #5.

A good doctor will be actively listening to you, without their eyes glazing over. Green flag #6.

Doctors who are present as type 1 will be okay with saying they know nothing about EDS, and a type 2 doctor will show an understanding of how

EDS can affect their medical specialty. Green flag #7.

If needed, the doctor is willing to educate themselves on EDS and what you experience. Green flag #8.

A doctor who will brainstorm with you if needed on treatment options is green flag #9.

A doctor willing to take home your stuff if the allotted time wasn't enough, AKA what I call "homework," is green flag #10.

A doctor who gives you referrals or names of doctors they have worked with. Green flag #11.

Making enough time for you to be thorough is green flag #12.

As you meet new doctors to consider working with, you will be able to tell if you are a good match or if they just are not the doctor for you.

## DISLOCATIONS AND CHIROPRACTORS

Many of us with EDS have been warned and advised against seeing a chiropractor because they can really cause a lot of irreparable damage if they don't know what they are doing. Unfortunately, we have laxity of our joints and a high chance of dislocations and partial dislocations, also called subluxations. These dislocations can't be treated by going to the emergency room even though the symptoms and pain can be severe and very debilitating. My problem joints are my ribs, and when they get out of place, it's never just one; it's always two or three. Having a rib out of place is an extremely sharp, stabbing pain that also makes it impossible to breathe.

Seeing a chiropractor doesn't have to be a negative thing. Having a chiropractor on your team of doctors is a necessity due to the sheer number of dislocations and the dislocation of the larger joints at the shoulder and hip. The ribs are difficult to put back into socket by yourself and having your vertebrae out of line can be very uncomfortable, causing some nerve pain if one becomes pinched. The chiropractor you want to go with is going to be the one who is willing to learn about or research EDS so that they are aware of the possibilities of potential injury and long-term consequences of being too rough when handling your skeleton. Chiropractors have tools they can use that are not only gentler but also react faster than your body. This works in being able to adjust the area before your body has the time to contract a muscle or become tense in anticipation. Another great thing about chiropractors is their use of KT tape. KT tape is a thick, stretchy fabric tape that they can use on parts of your body where it would otherwise be difficult or awkward to wear a brace. This tape can help support your ligaments and tendons like any kind of brace would while being much more flexible and comfortable to wear.

## MY TEAM OF DOCTORS

I have been with my PCP, Doctor Morrill, for twelve or thirteen years. He was with me in the beginning and has been with me since my open-heart surgery in 2011. He has seen first-hand my horrors during the undiagnosed period. Dr. Morrill is a type 1 doctor, and a PCP is a much needed ally. He's helped me manage a laundry list of medications, including medications for nausea, my thyroid, my non-narcotic pain meds for nerve pain, and my myofascial pain, to name just a few. He treats my shingles when I break out, and gives me the good cough medication if I need it. He helps with my respiratory needs by ordering my $O^2$, inhaler, and nebulizer machine. The best things he does for me is to help sign and send forms for several disability needs and requirements.

In my team of doctors, Dr. B, my functional medicine doctor, gave me my life back and not one where I'm doomed to my bed forever. Dr. Richards and Kelsie, my dysautonomia doctor and his right-hand lady, are helping to prevent my fainting episodes. Dr. Richards is always thinking outside the box and directs me where to go, and most of his referrals are to doctors he's worked with or to others who I know will treat me with respect and take me at my word.

I've been seeing my pain management doctor for twelve years. Dr. Garg has never dismissed the possibility that I have an incredibly painful disorder. Dr. Kira Law and Dr. Fred Wittleder were two of the doctors who were a part of the pain management clinic. These two made every appointment a lot of fun. It's one of the only appointments that I enjoyed, rather than dreading to have to leave the house. Dr. Fred always had the best shirts, it was just his thing. He told me once that only 10 percent of people who have medical insurance are the ones who *need* it. He told me that I was probably in the top 1 percent of this 10 percent of people. He was also familiar with EDS and he's the reason I was able to find the pain relief I needed. Sadly, he had an opportunity elsewhere, and with his departure, I've had the privilege of seeing Dr. Kira who learned from Dr. Fred, and she understood that I was one of Dr. Fred's special little people. In a way, I was the "problem child." I couldn't agree more, since it's rare that when I come in, I have nothing new to report.

I also have a fantastic psychiatrist, Dr. Tueller. I can tell you that from the first appointment, he didn't dismiss my EDS. There are endless benefits of finding a doctor who can figure out the best cocktail of antidepressants and anxiety meds that will work for you. Another thing that has been invaluable to me is the fact that he understands the hardships of having a chronic illness and the struggles we encounter. Personal validation goes a long way when looking for mental health help. I recommend you find someone who specializes in chronic illness.

There are also many reasons to see a psychologist as well. This disorder will affect everyone in your household, not just you personally. Many times, during high stress times or life-changing events, my husband and I have gone to Dr. Adams to help us work through the trauma and help us remember that we are in this together and that we are both fighting the same thing.

When I figured out that I was going to need a more aggressive plan to get nutrition or I wouldn't be around much longer to worry about it at all, I went to the local nutritionist who was feeding a lot of my local friends. I was out of options. Unfortunately, my insurance didn't cover Dr. Kinikini. She told me to get a tube, then figure out what to put down it. She helped me get in touch with a nutritionist who was really good with complex cases, named Marysa. I was reacting to every formula I tried. Marysa was ready to piece together a formula specific to me. It'll always take a few tries before you will find the right one. Marysa has been my nutritionist for several years. After eight years of starving, she was the only one who has been able to get me the only nutrition that my body has reacted well to, helping my body function at its best. There was no way I was ever going to work with any other nutritionist, not when Marysa understands how fickle my body is with all the ups and downs of water weight.

The newest doctor I've added is a fabulous cardiologist. Dr. Hancock is responsible for my longest run of being hospital free – close to a year. It's been the longest period being hospital free since I got sick with sepsis.

## MEDICAL SCREENING

From the time I was diagnosed with EDS, it only took me four years to find most of my team of doctors. Four years for a great team isn't horribly long. The hardest time we have is when we don't know what we are looking for. A lot of us don't even know that we need to be looking for anything. These are the hardest years to go through until someone happens to notice you've got EDS.

Having to go through years and years of medical

neglect, medical gaslighting, and physical and mental anguish is traumatic. It is all preventable and I came up with a realistic and easily implemented solution. There is no reason that pediatric doctors couldn't have two to three minutes of their time with each child at their yearly well child check-up to implement EDS screenings on all children five to eighteen years of age. Not only would this prevent them living a very difficult life, but the doctors would also learn about prevention.

I didn't know about EDS while my children were growing up. They both had a time in their lives when they had problems with their hips. The doctors knew there was an issue but were never able to tell me why. When my son, Logan, was only five years old, he began having enough pain in his hips that I noticed he had started walking slower, he wasn't climbing up the steps to go down the slide at the park, and he was walking with an odd gait. When he started having pain, I took him to our neighbor, who was a chiropractor.

Logan had always been a skinny kid and was always lower than the normal range for children his age and height. His body wasn't strong enough to holds his hips into place for very long. We took him to see the chiropractor once a week for about eighteen months before his little legs were strong enough to keep his hip joints in socket correctly. He also had severe "growing pains" in his knees his entire childhood. We had taken him to Primary Children's Hospital in Salt Lake City, Utah to get his knees checked and we were told that there was nothing wrong.

A little while later, we ended up going back to the chiropractor when I was getting transcutaneous electrical nerve stimulation (TENS) therapy. The kids were nine and eleven when we were told that they had flat feet, which could cause some structural problems. I remember the kids looked forward to seeing the chiropractor every week because they got taped up all over their backs, knees, and feet with KT tape. Most likely, their flat feet caused the structural issues later in life, that manifested as Logan's knee pain and later Morgan's hip problems.

Morgan started to have problems with her hips when she was fourteen. We had just moved from North Ogden to Syracuse, and she had a really hard time getting around the school. This school was more than twice the size of the last school and it was two stories. We had gotten special permission for her to use an elevator key. We had her looked at by a physical therapist and Morgan was told that she would probably be in PT once a week for four to six months. Still not knowing there was an underlying cause of EDS, she was given a time frame that was for a healthy bodied person. Morgan ended up going to PT weekly for about eighteen months as well, but her hip problems were a structural problem.

Medical screenings early on in childhood would not only aid the children but would also help get adults diagnosed with EDS much earlier in life. Since EDS is genetic, parents who have children diagnosed with EDS should also be tested, or at the very least know what to look for, especially if they have already started looking for answers.

## MEDICAL ETHICAL DELUSION

There is an ethical delusion when it comes to doctors in general, more especially when it comes to patients with rare, difficult, and complex cases. Everyone just expects that doctors become doctors to help people. Television shows like *The Good Doctor*, *New Amsterdam*, and *House* are all based on exceptional doctors going to great lengths to solve and get the patients the treatments they need; sometimes the patient was in a critical life or death situation.

We all assume that when you go to the doctor, not only will you get the help you need, but also the doctor will do whatever it takes to help, even if they don't know or have the answer. We all think that if the doctor doesn't have an answer or doesn't understand the illness, that they will do the research and refer us to other doctors.

Unfortunately, it's not realistic view of the medical field. Think of every job you have had – there is always some co-worker who enjoys being a jerk. It is no different in medicine. When you are treated badly, there is no manager you can talk to who can help get you what you came for. This makes it all the more important to find a doctor who is the right fit for you. You can guarantee if a doctor is mistreating you, other patients are being mistreated too. This makes it our responsibility to speak up, even if all we can do is leave a negative review.

## MEDICATION STIGMA

At one point in time, every single one of us has been treated in a negative way once a provider sees the long list of medications we take or sees that we are currently taking any narcotic pain medication. This bad treatment puts a feeling of pressure on us to reduce the amount of medication we take, or worse, puts on the pressure to quit taking pain medication. So many sufferers of chronic illness are made to feel like something is wrong with them for needing so many medications just to get through the day. Other times, our symptoms aren't taken seriously and blamed on the long-term opioid use. I've heard so many stories of people weaning themselves off medications or stopping some cold turkey, all based on how some doctors decide that stopping all these medications will be the answer to all our problems. When in fact, this is the opposite of the case; stopping medication abruptly can cause major short and long-term side effects and in rare cases, can result in death. I wouldn't recommend quitting any medication without a serious discussion with the doctor who has prescribed you the medication and not with any other doctor. The doctor has reasons why they prescribed you the medication because they believe the benefit of taking that medication outweighs any possible negative side effects.

The most difficult doctors to see are the ones who narrow in only on the fact that you are on opioids. It's also looked on negatively by doctors when you know exactly the medication and the dosage that works best for you when you are in pain that is too severe for you to manage at home. With the levels of pain that we experience with EDS and the countless times we are at the emergency room, it's not that hard to remember what works best for your body. My worst experience with medication stigmas was during a hospital stay when the doctor team that was treating me withheld one of the medications that was on my home list of medications. I even verified the medication list with several nurses and with the pharmacy. I've had a lot of questions about two of my medications that essentially are for the treatment of nerve pain. Only one of these medications is very effective in treating severe anxiety, which is why I am on both medications. The doctors on my team have approved me using both medications since both are at a very low dosage and each one treats a different condition. When I confronted the doctors about withholding my medication, they acted like they hadn't known I needed both. They were more than willing to allow me to take both medications, but they were unwilling to help me get my pain back under a manageable limit. Since they refused to help me quickly recover from the situation that they created, I signed an AMA and left the hospital. After such treatment, I'd rather die at home than ever be hospitalized at that specific hospital again.

## MUNCHAUSEN SYNDROME BY PROXY

I also have wondered about the kids who have Munchausen syndrome by proxy. The Gypsy Rose case has been the most publicly known case of this syndrome in the recent past. It makes no sense to me why these kids are put on medications, undergo medical procedures, and even have feeding tubes placed, when they obviously don't have the test results needed for these procedures. It's horrific that these kids are put through so much, yet people with real symptoms who face life-or-death situations don't get help and are treated so poorly by doctors.

Since we live in the real world, not the world that T.V. portrays, this is a real and huge problem. Not all doctors are this way. There are amazing doctors out there who will treat you and provide what they can, and when they run into the unknown, they will help you search for what you need. We need more of these types of doctors and you just need to find the one who will do this for you. It makes all the difference on your EDS journey.

# 8

# PREVENTION AND TREATMENT

One of the cardiologists that treated me while I was hospitalized gave me an information packet. It had a list of symptoms labeled: "Green: you're ok;" "Yellow: call your doctor or check in with your PCP," and last was a list of reasons for going to the ER, labeled "Red." The doctor was really cocky telling me the information was written by some "really smart group of cardiologists" and it was extremely important to follow those guidelines. I took a quick look, then told the doctor if I followed that green, yellow, and red list of theirs that I'd never be able to leave the ER! I'm not sure if that humbled her or not, but I could tell I hit a nerve and there was nothing she could say about it.

## NEVER GETTING BETTER

One of the most difficult things I get asked is, "So when are you getting better?" The honest truth is *never*! Your body will always make bad collagen; you can't fix that. All your muscles, bones, organs, and tissues are much weaker than they should be. It's so weak that our bodies will start to fall apart just after basic functioning. We are always getting something fixed, something that bypasses the problem of not functioning normally, and we need constant "upkeep" to keep us functioning as well as we can. There are very few things we can work on to get "better." We try adjustments, massages, fascia work, energy healing, acupuncture, cryotherapy, oxygen therapy, meditation, medications, and supplements, anything to help the eternal relentless fatigue and our pain.

I have always said that we won't get better, it just gets different. Some people take longer to change but there are so many things we have to deal with changing day-to-day and even hour-to-hour. Facing autoimmune issues and doing fascia work and mental therapy are just treatments that take time. On top of it, we have joint pain, weakness, subluxations, neck problems, and even need braces, slings, or mobility aids like a wheelchair, walker, crutches, etc., and this changes every day.

Finally, you have body temperature, heart rate, blood pressure, and autoimmune system issues that can change hour to hour. Besides all the "normal" EDS symptoms, I have several symptoms that are specific to me. In actuality, it took me twenty years to find answers and discover what worked best for me. Dealing with all the unknowns made it all the more confusing, frustrating, and extremely difficult to live with.

The first major complication was the birth defect on my heart that wasn't found until I was thirty. During both of my pregnancies, at age twenty-one and twenty-three, I experienced a lot of stroke symptoms, making my second pregnancy especially awful.

Still, when no cause was found, my case was quickly brushed under the rug and it was just suddenly over. When I had my heart surgery, I found out I was without 17 percent of my blood supply for thirty years. Then almost ten years later I learned of POTS and how our blood pools at the bottom of our bodies. Then add in the pregnancy, where blood supply is targeted at the baby. The combination of all three were the cause of my stroke symptoms.

Another issue that was specific to me was the gain and retention of obscene amounts of water.

Nothing I did with any diet and exercises helped me lose a pound. I could gain thirty or forty pounds overnight, and upward of seventy-five pounds with a one week stay in a hospital.

Once, I gained 100 lbs. in four days after a coma from sepsis where my liver and kidneys were in full failure. This prevented my body from voiding any urine, compounding the water again. I have two doctors I saw one month before and one month after this hospitalization and medical records to prove this. I couldn't lose weight and when I started having pain with eating, this water weight made it take eight years before I found a doctor who believed that I wasn't able to eat.

In 2013, right after my two ectopic pregnancies, I sought out a doctor for help. I found a liver doctor who told me I had to have weight loss surgery, or I'd be on the liver transplant list within five years. Since I didn't know any better, I spent one year jumping through all the hoops necessary for weight loss surgery and trying to come up with $20,000 for the surgery, since it wasn't covered by insurance. I was told twice a month at doctors' appointments to eat more. No one listened.

I had the surgery in the spring of 2014; after six months, I had still only lost thirty-five pounds. The doctor yelled at me that I wasn't following the plan, and he was mad he even did the surgery. That appointment, he happened to bring in four other doctors. He yelled at me again about not following the plan. I yelled back that from day one over a year ago I wasn't able to follow the plan because you'd have to eat in order to do that. Then my husband stood up and yelled at him saying he was an idiotic jerk, and that he wasn't listening.

The doctor finally said he'd do an endoscope to measure my stomach and had a cardiologist to measure the pressure in my heart. The endoscopy came first, and he found my stomach was the correct size, but I had developed a hernia. The cardiologist told him she was 95 percent sure I had Cushing's disorder, when water is stored in fat and very hard to lose. The surgeon was finally proven wrong.

He also told me the hernia was because of the stomach surgery and that in fifteen years, I was his only patient to get a hernia. I saw an endocrinologist who said I didn't have Cushing's based on this test. Again, I was told it wasn't Cushing's, good luck!

In 2017, I saw an autoimmune doctor who was also a functional medicine doctor. After eighteen months of treatments and a six-hour roundtrip each visit, I went from being in a lot of pain and bedridden to having a part-time job delivering packages, and able to help care for my husband's grandparents and my family. This doctor balanced my hormones and got me healthy enough to no longer be borderline lupus and to prevent possible colon cancer in less than five years.

It wasn't until 2018 that I found a dysautonomia doctor who finally treated my fluid overload with diuretics. Even though we could fight the retention, I was yo-yoing every time I was in the hospital. A pediatric psychiatrist came out and said that adults raised in narcissist families would constantly be in fight or flight and could develop Cushing's as an adult. When I was diagnosed with EDS, I learned that it's common for people with genetic disorders to begin third-spacing water, which occurs when water is stored in space between cells.

As a result, I was so extremely bumpy and had small areas with pockets of fluid. It wasn't until after my back surgery in 2022 when I was finally treated in hospital for the fluid and was on high IV diuretics and peed, losing 85 lbs. in three weeks. With my feeding tube and the right formula (that took two years to find), I found I could finally prevent my body from increasing fluid retention. To lose water, I need to eat as little as possible – 95 percent of my nutrition comes from my tube feeds and I could finally, slowly lose the water and keep it off.

Multiple doctors tried to get me to stop tube

feeds, telling me it wasn't food and only food was going to help me get better. *None* of them understood that my body can't break down and process food orally – it was like poison to my body, the way I reacted to foods. With the feeding tube bypassing my stomach, preventing the pancreas from making enzymes to digest food, in turn, I no longer had pain while on tube food. After only two days of no tube feed and still little food orally, I gained 30 lbs. back and was then discharged.

A few months later, I was in and out of the hospital with fluid overload so badly my lungs needed ICU levels of oxygen and my heart had one valve that start to leak – only making the water retention worse. After six months of in and out of the hospital, I finally found a cardiologist who knew what to do and has helped me lose enough water that it allowed my heart ventricle to close. I have no more leaking and my oxygen needs are very low. I haven't been hospitalized for nine months.

Because of my overworked heart due to my heart defect, my water gain began six months after my first baby was born twenty-two years ago – that's a long journey with water retention! Most people with EDS don't have these issues, but all the issues played off each other and made everything worse.

If I'd had a doctor believe me earlier, or had I

been diagnosed earlier in my life, my life would have been so much easier and much less painful. Awareness and early diagnosis are our best shots at the healthiest life we can have with EDS. Since it is not curable, all we can do for our bodies is prevention and treating the symptoms we have.

## PREVENTION

Prevention may not sound like a great care plan or plan of action for your EDS. I thought so too, at first, but living with EDS and being extremely symptomatic has shown me how essential prevention really is. Our bodies can heal slowly but every time we have surgery or an injury, there will be a lot of scar tissue that could possibly cause irreparable damage; our natural wear and tear will happen faster, and we can make it worse.

I watched a friend have the spine-fusing surgery for cranial cervical instability (CCI) and it was brutal. She had to shave most of her hair off and the incision started on the top of her head, going down her back almost to her waist. Standard recovery is one year. She was also brave enough to do all she could to ensure the surgery's success. She went across the country to one of the only five surgeons who will perform this procedure. Then, to ensure that her nerves weren't affected, she was conscious during the procedure. Most people will also have a secondary surgery to untether their spine. Being such an invasive surgery, she had a lot of cognitive problems – she had no attention span, no memory, and she had difficulty having a five-minute conversation.

I experienced a lot of these symptoms too every time I went septic or into a coma. It's just as difficult to recover mentally as it is to recover physically. Knowing what I know now, I think back to those times I went on some kind of roller coaster or amusement park ride, and I can't believe that I haven't caused more damage to my neck. This puts prevention in clear perspective.

Once I was told that I was fragile. I was being released from the hospital again after one of my many near-death experiences. The doctor, Dr. Abe Care, was definitely the reason I came out of that alive. I don't remember a lot of that conversation. What I do remember is that he told me I was fragile, and he asked me, "What would the sign say if I put one on your back?"

It's very vivid in my mind, but I pictured myself wearing a huge sandwich board with "fragile" on the front. All I could think about were "fragile" labels on packages. I told him I would write "handle with care." That moment was my favorite time in a hospital. Later, I realized how true and fitting the phase was. We can easily dislocate one of our largest joints just by being picked up and moved if it isn't done right. In that case, we are extremely fragile and should be handled with great care.

It's impossible to prevent everything all the time, but it is vital enough to always think about prevention. Prevention is all the more important because doctors are reluctant to properly treat an injury and will only do so after sufficient time to heal naturally, which can take extremely long with EDS. Even then, you might have further testing done, usually an MRI, just to prove that the type of injury isn't something that can heal on its own – that surgery or a procedure is the only way to actually heal.

I rolled my ankle, which I walked on for two weeks before I knew it wasn't just a sprain. The urgent care did an X-ray and sent me to see an orthopedic doctor, who explained that I had an avulsion fracture, where the tendon tore off a chunk of bone at the attachment. I was put in a boot for six months to heal. Being my right ankle, I wasn't able to drive the whole time. Six months later, my ankle still hadn't healed and then the surgeon ordered an MRI. Along with the avulsion fracture, I had torn all my tendons, and no amount of physical therapy was going to fix that. I needed an unexpected surgery to reattach every tendon and then was still in a boot for eight more months for full recovery in another boot. Had the doctor run the test in the first place, it would have saved me eight months of being in a boot and I could have been able to drive and not be so isolated. Had I known I had EDS at the time, I would have tried to have the MRI done right after the injury.

## TREATMENTS

Our only other option is to treat the symptoms we have. Our bodies aren't going to be able to always function correctly and your body cannot return to proper functioning.

There is no "fix." Once your body stops working efficiently and effectively, it will always function this way. Sometimes, we can replace what doesn't work. I had an infection of osteomyelitis, an infection inside the bone. The infection wasn't found for about six months. I had passed out once and hit the wall, not even that hard, but I ended up fracturing two of my vertebrae. With the initial fracture, I was put into a coma due to pain. Then I spent two months in a physical therapy place, and a few weeks later, I was back in the ER for back pain. I had to be heavily medicated to get in the MRI machine, which showed that the two previous fractured vertebrae were now just in shards, basically eviscerated. After replacing my spine from my armpits down to hips with titanium hardware, I spent a little more than a month in another medically-induced coma to allow my body time to heal with enough pain control. Even replacing my spine, I had a lot of physical therapy and residual pain.

Unfortunately, we can't replace everything in our bodies that stop working. For most things, there isn't even a fix at all. Instead, we receive a maintenance plan with so many complications there is no way to help you feel better. We might have fifty small things that help you feel only 1 percent better. Also, not everything works the same for everyone. We all must find our own personal cocktail of treatments. Some treatments may only be necessary for a small amount of time, while others you will need for the rest of your life.

There are so many treatment options out there. Some common ones are: IV saline, careful physical therapy, massage, acupuncture, ice baths, fascia work, TENS units, supplements, prolotherapy, trigger point injections, medications, breathing and swallowing exercises, nutrition, chiropractic work, energy work, hot stones, talk therapy, EDMR, functional medicine, continual appointments with cardiologist, GI treatments, neurology work, or orthopedic work.

At some point, you may need your doctors to work together to get you the care you need, which is why you should never stop looking until you find the right doctors you need. Treat yourself like you would want someone you love to be treated. If there is anything having EDS has taught me, it is to be more compassionate to others. Everybody is fighting their own invisible illness; we never know what someone else is going through or how difficult it is for them.

We also need to be more compassionate with ourselves. It's okay that we aren't okay and it's okay that we'll never be okay the same way we used to be, because our lives will never be like they used to be. We can't be so hard on ourselves. Most likely, you've always been hard on yourself, trying to keep to a certain standard. It's okay that we didn't return that phone call right away. It's okay that you need more rest and have less time to socialize. It's okay that you can't do all the things you are used to doing. The best treatment that is out there for you is how you treat yourself. We are our own best treatment.

# HOSPITALIZATIONS

Most of the time, other people don't understand that you can only get better to some extent. These disorders are degenerative, meaning your body will degrade more over time as you get older. Since the condition giving you the most symptoms and the most problems will only get worse, there's only so much that you can do other than to treat the symptoms you have and use whatever you can to prevent injuries and illness. As symptoms arise, there are some things that will affect us for the rest of our lives. Imagine that as our bodies continue to regenerate and heal, all that is available to rebuild are building blocks full of termites. We won't be able to all of a sudden have access to healthy wood.

I have several conditions that make it very difficult to leave the house. In order to go out, I have to bring everything with me. I thought that I packed the entire house in my diaper bag when I had young kids, but hey, that was nothing compared to a day out, whether it be for doctors' appointments, to see friends, or go out to a movie. What does matter is when my tube feeds need to run to be completed by the end of the day. I also have medications that are on a schedule, and these have to be easily taken with me.

When you have a disorder that doesn't have one single specialist, *you* have to be the specialist. You must manage all the symptoms you have. This means that along with your regular check in with your primary care physician (PCP), you will also be seeing any number and combination of doctors. You can have a cardiologist, a neurologist, an oncologist, a gastroenterologist, and a pain doctor. And you will be required to see them on a regular basis. For some things, you may even need surgery. Depending on the doctor and the type of surgery performed, you may or may not have to stay at the hospital for up to a week after initial surgery.

## HOME HOSPITAL

I went septic in November 2019 and stayed in the hospital for almost six weeks. Then exactly a year later to the date of the last admission, I was back in the hospital for pneumonia and sepsis. Then of the next thirty-three months, I spend twenty-six of them inpatient at hospitals. I had to be in whichever hospital had an open bed because it was still in the height of the COVID pandemic. I was in and out a few weeks; one time I was discharged, only to be admitted again less than twenty-four hours later. With all my years of bronchitis and now fighting pneumonia, I've also been having complications with my lungs. Having been in hospitals so long, you start to make and cultivate relationships and friendships with any of the staff, housekeepers, nurses, certified nursing assistants (CNAs), or doctors. I have been lucky enough that the local EMTs and emergency doctors and nurses know me and my condition, so once I am seen, I am usually moved out pretty quickly and sent up to a regular hospital room. This hospital is the one I call a "home hospital," where the doctors, nurses, and other hospital staff get to know you and believe your condition.

One time, when I was bedridden in a hospital stay and in incredible pain after having chest tubes ripped out by a doctor, the CNA took the time to French braid my hair, even though I'd be in bed all day. It is little kind gestures like this that make all the difference during a hospitalization.

I know that if I go to my home hospital, it's a safe place for me and I am treated with decency and respect. There was a time when I struggled with bacterial pneumonia that then became septic, because my body didn't have a healthy immune system that functioned correctly at the time. A perfect example is that after about six months of getting two vaccine shots for COVID, I still had no COVID antibodies in my blood. This showed that my immune system was not working as it should, which would have normally build antibodies to prevent me from contracting the full virus.

I have repeated this cycle ten to twelve times and the last few times that I have been to the hospital, I walked in saying that it hurt to breathe. I was immediately taken back to a room and saw the doctor on staff quickly. Right away, blood was taken, I had an X-ray of my chest, was put on heavy IV antibiotics, and was told my room was being prepared. Normally, nothing is done until test results have confirmed a diagnosis. Since most of the doctors know me or have seen my chart, I'm pretty sure there were notes in my file to not let me go home.

This has not only shortened my recovery, but also saved my life on several occasions. Once I was already unconscious when I arrived at the hospital via ambulance. I was in a coma and my family wasn't sure if I was going to come out of it. My kidneys and liver had gone into failure. Without my kidneys and liver to filter and pump fluids to my bladder, my fluid retention compounded exponentially. The quick action by the doctors and nurses at my home hospital helped with my recovery.

I do not get as sick as I have been. Before last year, I had gotten severely septic, and I was picked up by an ambulance and taken to the hospital where I arrived unconscious and was in a coma for four days. When I first opened my eyes, the first thing I saw was my daughter. She is the most loving, kind, and caring person, with the nature above and beyond any other teenager that I know. She let me know that I had been out for four days and that doctors were not sure if I was going to make it. I ended up in congestive heart failure. I had to relearn everything. I couldn't even feed myself; they had to have someone literally feed me to get the small amount of nutrition that I try to meet daily. I couldn't even go use the restroom by myself. I had to rely on all the nurses and CNAs to take care of me. It was an interesting experience not being able to feel my body; it was also one of the most difficult things to overcome.

## SEPSIS AND POST-SEPSIS SYNDROME

Many people don't understand what exactly sepsis is or how intensely it affects our bodies., especially our brains. Sepsis isn't something you can just catch like the cold or some other contagion. Sepsis occurs when you have some kind of bacteria or infection that is resistant to antibiotics. In other words, sepsis is your body's reaction to bacteria. If you do not get enough antibiotics to kill the bacteria that is in your body, then your body begins to shut down. We knew that my liver and kidneys were failing, and my heart was in beginning stages of failure. Sepsis follows a pattern, and it shuts down each organ, one at a time, so after my liver and kidneys, my heart started to shut down, and this was why everybody was worried that I would not wake up. I had gotten really close to the point of no return.

When this happens, your brain can be affected. There is a condition called post-sepsis syndrome. I mentioned that when first coming out of a coma, I must relearn everything, even the most basic things. This is all because of post-sepsis syndrome. I remember the first two days after I woke up. I spent both days in bed because I could not move anything in my body. I could not physically get out of bed. Then, when I could get out, I had to relearn my body signals for when I needed to use the restroom. This was the worst, because for a little while, I was a little off and didn't make it on time.

Along with learning how to use and take care of my body, I hallucinated often and the hallucinations made no sense. I watched an enormous, beautiful dragon fly through the wall. I kept seeing a dead rat in my bathroom, and it only went away after I put the garbage can next to the wall. There was a three-day period when my daughter's face had really gorgeous tribal make-up on. Hallucinations are not all pleasant to look at, and some are scary. While driving on the freeway, I saw a huge truck coming at me head-on. I was so lucky that I didn't accidentally kill myself when I swerved and pulled off to the side of the freeway onto the shoulder. Then there are hallucinations that are called "all immersive," meaning 100 percent of what you see, surrounding you 360 degrees, are all a part of the hallucination. I had one of these types last for three hours.

These hallucinations can also be auditory, since EDS can affect your eardrum, thereby affecting what you hear. I'd start conversations in my head and eventually started to speak my side of the conversation out loud, even when no one else was in the room with me. I didn't have an attention span. I couldn't read, watch T.V., play a game, or hold a conversation over five minutes. I'd forget what I was saying right in the middle of a sentence. My favorite times were when I'd ask my husband a question and immediately after he gave his answer, I'd say "Great! Now what did I ask you?" I couldn't keep ahold of anything in my hands for even a few seconds. The most hilarious times were when I'd take a sip of water and immediately spit it out all over my husband, as well as the times when I would dump an entire mug of ice-cold water out all over myself. By chance, I had one of these times recorded, and it looked like I did it on purpose, except looking at the surprise on my face proved otherwise. Other neurological issues from post-sepsis syndrome are loss of muscle control, slurred speech, and fine motor issues.

One of these times after I'd gotten home, I had woken from a nap, and I got up to use the restroom. With the extra 100 lbs, my body was not equipped to hold the weight. I like to use this analogy. It would be like a person putting on a backpack and then every day, you put in one block of butter. As you gain the extra weight very slowly, your body knows how to handle itself and there are lot of things you are more capable of doing with the extra weight. Whereas my situation would be just going to the fridge and picking up ten gallons of milk and deciding, now I'm off for the day, and I can't put the gallons down. This puts a strain on your body, your bones, your muscles, and it induces more swelling. With the weight distribution and my body having the issues it does, my body first stores it in my fat, then what is called "third spacing" in my body. When I got up to go to the restroom, I kind of leaned a little forward to start walking, and I just kept picking up momentum. I could not stop myself! I used the wall to stop and just let myself hit it so that I didn't fall down. I was in a strange position, and between my husband and myself, I could not get off the floor. I was lucky though that my dysautonomia doctor gave me a diuretic other than Lasix. My body lost 10 lbs. per week for about nine weeks. As I was losing the weight, I was able to get around more easily and be able to do more activities.

## COGNITIVE IMPAIRMENT

I remember the second time going septic was the worst. I was having conversations in my head, and I believed that they were real. Eventually, I got to a point where I started talking to these people out loud. That's when I realized that this is what I had been doing all this time and it really left me confused.

It took me a long time to have more than a two- or

three-minute conversation, since I could not focus enough to get my words out. I kept dropping everything, which is something that I did prior to going septic, but afterwards, what I was dropping was seriously ten times worse than it was before. It took months until I could focus, concentrate, and function on my own.

## HOSPITALIZATION TIPS

Even now as I write this, I am in the hospital again, having just been released three weeks ago. This time, it appears that in one of my falls, I fractured one of the vertebrae in my spine and am in some of the worst pain I have ever felt in my life. Most of the time you are left alone and do not have a lot of visitors.

For all of us, hospitalizations are one of the necessary evils of EDS. I've heard many people with EDS tell me that their hospitalizations might have very well been hell. I've had many of my own stories of horrors happening which I can't share just yet. I've done it all – like taken off my monitors and taken off my gown (that seemed to get a word with a doctor). I have also signed an AMA (against medical advice) demanding to leave because the treatment was that awful.

I've been in enough ERs to know that it's hit and

miss for how long it takes. I always remember to bring one or two doses of my pain meds with me, in case I'm in the waiting area for several hours. Once you're hospitalized, the only meds you have access to are the ones given to you by nurses based on doctors' orders. Most of the time, it's an unexpected hospitalization, so having a small overnight bag can help, especially if you don't live close to a hospital. This gives family a bit of extra time before they have to come back with the rest of the things you need. I always like to have a few comforts from home, usually PJ bottoms, socks, and my Minky blanket, or else I freeze. Finally, bring something to do, like puzzle books, books to read, or anything you do at home to pass the time.

# 10

## WHAT'S REALLY GOING ON IN YOUR HEAD

At some point in time, most of us have been told that these symptoms and conditions are all in our head or that we have some psychiatric or conversion disorders. Now, we know that our symptoms and experiences are very real, and we have medical needs. Most of us have mental issues, but they are not our symptoms. Everyone with chronic illness will have some degree of depression, whether it be depression or a severe mood disorder due to our chronic illness. A lot of the time, we feel isolated because of our symptoms, pain, and all the time spent in doctors' offices. We stay home from social gatherings, meetings, and fun nights out because our pain makes involvement physically and energetically impossible.

Most people will give you recommendations,

saying things like, "Try this new therapy," or "So and so worked for my friend," and this may be stuff we really don't want to hear. We may need support from friends and want to talk to them, but we really don't enjoy having sessions with them where they offer all of their ideas as to what we should do to get better. The thing is, they don't know better – you and only you are the expert because you're the one who lives through it every day.

Unfortunately, our pain and low energy keep us close to home and friends or family may stop hearing from you or stop trying to get ahold of you. Perhaps you are saying no to too many things, so the other person eventually, no matter how long you've been friends, just stays away for good. I myself have gone through this and have had to cut ties with someone that I had known for twenty years. We went through many things together, and at one point, we thought that there wasn't anything that could break our friendship. But apparently, that break came when my current disorder had left me disabled. These things – pain, fatigue, and isolation when cut off from family and friends – can cause depression.

## PTSD FROM TRAUMA

We can also get anxiety due to our POTS when it's left us out of breath or when we pass out, very disoriented, confused, and panicked. When it happens to you repeatedly, it just gets scarier because it's happening over and over. Only when we have treatment to care for our various degrees of severity does this panic go away.

We also experience a lot of PTSD. This comes from trauma that you've experienced, and later, other situations or similar situations can trigger those memories. How traumatic those experiences were will determine the severity of your PTSD. PTSD is a very difficult disorder to have. The past feelings will come on hard and fast, and sometimes seemingly out of nowhere. When these feelings hit you, it doesn't get better. It feels like you're experiencing the exact same trauma right then and there.

As far as I understand, there is no cure for PTSD. The best that you can do is learn how to handle and cope with your PTSD. In that moment right when it hits you, you will learn how to recognize symptoms and signs of your reaction. Eventually, you will know and feel the sequence and be able to determine the severity of the reaction. Once you feel you are in the moment, you need to just stop everything – stop your phone call, put down your pencil, or stop your conversation. You need to stop and then take a little bit of time to ground yourself in reality. There are many ways to do this, first by stopping everything that is going on, and then recognizing where you are. Sometimes it helps to look at a calendar and not the date. Anyway you get yourself grounded – it doesn't matter; all that matters is that you got grounded in your reality. You know what's going on right then and right there, and not in the same experience that created the trauma. This doesn't mean that PTSD goes away, because it doesn't. Instead, this will help tone down your reaction so that you can move on with what you were doing in the moment.

## PTSD AND MEDICAL PERSONNEL

At some point in time, most of us have been told that all our symptoms are not physical but psychological. We are written off as having conversion disorder, which is when something emotionally presents as physical symptoms or any number of psychiatric symptoms. One of the hardest things we all have to deal with is that EDS is an invisible illness, so we all look normal. We also look younger than we are because our connective tissues can't hold the wrinkles tightly. Most of the doctors I've seen haven't believed me. I've been called a liar, faker, attention seeker, and once I was even expelled from a clinic. I even received a letter saying they would no longer treat me for any reason forever because I was labeled a "drug seeker."

I had severely bruised my tailbone and was told to come back for a refill. At that appointment, I was grilled and pressured, all because I didn't make the appointment with the doctor who originally treated me. I had multiple children in my care along with my two kids. I showed up when I could. The doctor got in my face, yelling at me saying that I should have seen the previous doctor for continued treatment, even though the clinic was an urgent care. She told me that no one in this entire office would treat me and that she wasn't going to do me any favors, and then barked at me to leave.

I wouldn't be able to count all the doctors over the last twenty-three years who have not only tried to get me to think that my symptoms were psychosomatic, but also who have tried to convince me that there was something wrong with me, that my brain wasn't working correctly. I had a doctor demand I take a psych evaluation before he'd agree to treat me. I started to do another eval with their chosen psychology doctor. Since I knew I was going to die if

I didn't get nutrition, I met with the psychologist and in the first session, I was asked why I wanted a feeding tube. I told him I didn't *want* a feeding tube, but I needed one desperately. We talked for a few minutes because he wanted to have a preliminary conversation with me before the evaluation, which was to happen at an entirely separate appointment. He told me he knew that I had a sound mind, and I knew exactly what was going on with me. I'd decided not to return for the actual evaluation. I had already done ten previous psych evals, each with a different doctor, and all within the last eight years. If that wasn't enough confirmation for the doctor who had referred me, I wasn't doing any more work.

If you haven't been convinced that you have a bunch of mental illnesses that are the cause of your physical symptoms, then they try to convince you that you are making it up for some reason – maybe they say that your brain is sending wrong signals, that you are going crazy, you are losing your mind, or you are forgetting things. After so many times, you start to question yourself and your symptoms. With EDS diagnosed, it's hard to even know what you remember, and which parts are real.

When I first got my feeding tube and had the most nutrition I'd had in five or six years, my doctor treated me for a year. Everything went great until my port got infected and after three courses of antibiotics, I ended up going septic for my first time. As it was time to discharge, the doctor came up to my room. He sat in a chair, put his feet up on the desk and his hands behind his head, and leaned back. I don't remember a lot of what he said. What I do remember is when he pointed to his stomach and then pointed to his head, saying that my pain wasn't coming from my stomach and it was really coming from some psychiatric condition.

I was told that I was just fine. That the pain I had been feeling for the last six years was just in my head and I was going to go home and be just fine. I was told to eat whatever it is that I couldn't eat. Then he tried to gaslight me by saying that he believed in me, and that he had tried to help me with my gastrointestinal issues, but ultimately, there was nothing to treat.

He kept trying to tell me that he knew exactly what EDS is and it has nothing to do with bowels; therefore, he had no reason to treat me. He said he would treat me for the next ten days on an emergency basis only, but then I'd need another physician because I was far too complicated for him. I remember sitting in the chair with my husband next to me and I was just sobbing.

My husband, being who he is, looked up four

peer-reviewed journals that were published. Knowing that the doctor wouldn't stand by his word and honor my prescription, my husband emailed the doctor's practice the four journals that had been published on how EDS specifically affects the GI tract. I was devastated. I knew I was going back to being unable to eat food and get the nutrition I needed. I knew I'd be having so much less energy. My hair would continue to fall out and my nails wouldn't grow. The absolute worst thing was going to tell my kids what had happened. I knew that they would be devastated, upset, and angry – they didn't want to watch their mom die.

Eventually, my physical test results for malnutrition showed up. It took two years of trying all kinds of formula to find one that I didn't have an MCAS reaction to and another two years of actually getting the nutrition I needed before my body tested within a healthy range. My finger and toenails stopped growing, half of my hair fell out and started to turn prematurely grey. I had passed out once, and even though I didn't hit anything, I fractured my ankle. This is just a little of my experience, but enough to show how much impact these doctors have on our mental state. A very high percentage of people with EDS would rather stay home and deal with a medical emergency by themselves than go to another hospital or ER. Along with all the physical symptoms, these conditions cause some mental conditions. Having an illness that changes every aspect of your life, it only makes sense that it will affect you mentally. The most common mental illnesses that those with EDS experience are depression, anxiety, and PTSD.

THERAPY

We all need a psychologist or therapist for talk therapy, evidence-based therapy, EDMR, and other coping techniques. We also may need a psychiatrist to aid in our therapy and to prescribe the right medications to assist with our mood and assist in getting adequate sleep.

Getting better mental health takes time to work through all the trauma you have experienced. With newer ways to help you deal with depression, emotions, and trauma, it's worth looking into it. Prolotherapy and ketamine have been shown to help decrease pain. While pain may not be a mental illness, the less pain you are in, the more active you will be. This has a large impact on your life, because getting back to some of your regular activities and feeling like your old self does amazing things for depression. There are several therapies that have been approved for help with processing trauma – such as ketamine, evidence-based therapy, and EDMR – which work differently than talk therapy does.

The first step in the healing process is that you must decide that you want to heal and prepare yourself for the effort you need to put into it. Opening yourself up to healing means giving yourself permission to heal. Remember that the best things in life are rarely the easy things.

## SUPPORT IS VITAL

People don't understand why we are so exhausted and in pain all the time. Unfortunately, these are the symptoms and side effects of our condition. It feels like others can't comprehend why or how we can be easily confused, unfocused, and forgetful.

We know first-hand that life can be taken from us and that our health can decline quickly. We can't make it through these situations, and we must depend on others for help. We'll also need someone we trust who either knows everything about our condition or knows where to find that information. Since 2019, I have had to rely on my husband on too many trips to the hospital when I have arrived at the hospital unconscious. There have also been times when I passed out at home and had to call for reinforcements. The amazing first responders have been so kind to come pick me up and take me to the hospital on an ambulance ride.

It's very important that you have someone who has access to a list of your medications. These contain key bits of information, and the earlier and more detailed this information, the better. It can be necessary and can save your life. This detailed information will also help prevent any further complications, such as having an extreme allergic reaction to either a medication or the combination of medications that are contraindicated. It is also very important that you tell all medical professionals a little about your EDS. As zebras, our bodies are not as strong as those with a healthy body; we are set up for easy injuries and dislocations. There have been two to three times when the EMS team was not in sync with each other while carrying me on a tarp from my home to the gurney, leaving me with a few dislocated joints.

I can go from healthy looking and joyful to almost dead in just a few hours. My home hospital has had enough experience with me in the ER, ICU, medical / surgical, and telehealth to be familiar with my condition. From late 2019 and all through 2022, I had gotten pneumonia a total of twelve times. There were many times when I would arrive to the

ER, merely saying it hurt when I breathed, and I was immediately put on antibiotics and told that they were just waiting for a bed to open up. Once the room was open, I'd be taken to my room and admitted before any urine, blood, and imaging came back from the lab.

## CONTRAINDICATED LIFE

We need a lot of support to live our lives. I can no longer drive due to the side effects of sepsis, since I can get confused easily and have times when I hallucinate. Finding just the perfect balance of support and treatment takes a lot of trial and error. Each time we introduce anything new, there are always some kinds of consequences, such as anaphylaxis, rash, IBS, and varying degrees of brain fog.

I am also fluid overloaded and it's now affecting my heart. Everyone knows that for heart health, it's important to have a low sodium diet, but with POTS, it is recommended to have a high sodium diet. Another way to decrease POTS symptoms is to get IV fluids to help keep up our blood volume and prevent syncope and presyncope episodes. This is also in direct conflict with having too much fluid, causing fluid overload. Another impossible way of living is that standing up too fast and walking too quickly can cause syncope. Having weaker muscles and tissues as well as two pregnancies, I ended up with severe bladder prolapse. Once I feel the need to empty my bladder, I only have so long before I lose that control and have an outright accident. I only have so much time to get to the bathroom but have the very real possibility of passing out. Another thing we are told to do when dizzy is to put our head down between our knees. But since I have a PICC line, I need to keep my head above my heart. If I bend over too far or for too long, there's the possibility of dislodging my PICC line, which would need to be replaced to be able to use again. Having EDS and needing diuretics creates very high amounts of urine. With this disorder, there are a hundred different ways that your life is in constant contradiction with itself, which is exhausting.

My personal contradictions require me to juggle the following:

### *Salt*

- I need salt for dysautonomia.
- I need low salt for heart.

### *Peeing*

- Get up and walk too fast, I'll pass out.
- Walk too slow, and with my severe bladder prolapse, I pee everywhere.

### *Fluids*

- Too much fluid is bad for my heart.
- Not enough fluids and I pass out multiple times a day.

### *Bending over*

- You are supposed to put your head down when dizzy to prevent passing out.
- I can't put my head too low because I could dislodge my PICC line.

## LEAVE TOXICITY

We all have surprising experiences with family and friends who we thought were loved ones. The truth comes out when we get sick. This illness takes over your life, and unfortunately, when this happens, this is also the moment when you know who loves you and who is in it for the long haul, no matter how bad it gets. I have a terribly toxic family and have had to go no contact with my entire family. I met a friend, and we were each other's family; we raised our kids together, we played, cleaned, worked, shopped, and shared so many meals together. We were closer than sisters and saw each other daily for seventeen or eighteen years. I believed that she was going to always be there for me. We had experienced so much together, including extremely difficult times, scary times, and divorces. Up to this point, I had no doubt or thoughts of her not being part of my life.

And then, I went severely septic for my first time and was in the hospital for a month. She was around more, and then she wasn't, but it didn't negatively affect our friendship. Exactly one year later, I went septic again. This started my eight-month hospital stay and I went septic five times, had three massive bowel bleeds, and pancreatitis. I also had come very close to death six times and I had to get blood transfusions for a year, and was on oxygen for two years. It was all very traumatic. I thought this friend understood more than most because she was a nurse, so she had experience with sepsis. In the first eight months, I saw her twice and she didn't stay long. We set up a day to get together after I got home. I had this feeling that something was going to come up and that we wouldn't get together. She called me that morning and told me she couldn't come but it was for a good cause. Some co-worker needed someone to work their shift so they could attend a funeral. I understand wanting to help but I knew this was an escape for her. She came up with fifteen hours for an unknown co-worker but didn't have fifteen minutes for her best friend who almost died several times in a year.

## EDS HAS ITS OWN COMMUNITY

It's so important that we do what's best for us. We have too much to deal with. If you have friends or family who don't support you, don't believe you, or tell you that you are just being dramatic, it's harmful to keep these people in your life. We have serious problems and being around toxic people doesn't help; in fact, it hurts us. If you have people who are not around when you need them, when you've been there for them, it is extremely toxic. The situation is devastating and it's important to go no contact. After you do, you need to allow yourself to grieve these relationships.

There are others out there who do get it and understand what you have to deal with, from the difficult times to the everyday annoyances. I found better friends for me. We need so much support and I'm closer to these ladies then I ever was with anyone I cut out.

It is really difficult to be "forced" to spend a lot of time in isolation. If you are hospitalized for a short time, it can be quite a nice "vacation from life." If you are in hospitals for months upon months, it gets hard for people to continue to visit. Even if someone comes daily, you are still limited to the time your visitor has. Additionally, there are many rules around visitation, especially during the COVID pandemic, making isolation worse.

I know that I am a chatty patient. Most hospital staff tend to like me. Since you see these people more often than you do some of your friends and family, our nurses also are the ones we will need to have around to help you with the embarrassing moments. I have been in multiple comas, so I know what happens, and most likely, you will be in a diaper. I remember a five to ten second moment when I was woken up, thinking, "I'm a forty-year-old woman and I'm in a diaper!"

When you spend time with someone and share embarrassing moments or near-death experiences with them, you tend to make closer connections. Add that to how often you'll be spending your time at the same hospital, and I'm so glad that my home hospital is a small one. I have had almost every single nurse, multiple times. I have much respect for nurses and appreciate all they do for me, and I don't always get to tell them.

I approach everything with humor, and I did at some point decide to appoint myself as a professional patient – having spent twenty-three years seeing doctors and nurses more than anyone else. I knew I wasn't ever going to be capable enough to complete my degree and absolutely wouldn't be able to complete the doctorate I'd strived for. Knowing the odds of where I'd be (the hospital), I still wanted a career, so I gave myself one: a professional patient!

In this role, one of my responsibilities is that if the person I was working with was kind and always ready to help, that I was to make sure that they had a great time in my room. I always make sure everyone laughs and is entertained while with me. You won't believe how many of them have thanked me for making their job easier and they are always excited to work with me knowing that at least for part of the day, they would get the chance to laugh and joke around themselves.

I know I was one doctor's favorite patient. This

doctor is the reason I am alive and will always give me the care I need. I know there were times I'd go to the ER, and he did something, because even when I looked fine and was in a good mood, that doctor made sure I didn't leave. What he did for me helped in getting all the doctors to believe me and always take my condition seriously because I can really be in such a bad place within hours. I also know that there were several times when the hospital was filled to capacity, but I wasn't turned away; they found any room they could put me in. This doctor made a way to ensure I'd always get the care I need. Most likely, he has created something that will likely save my life in the future as well.

## EDS BOOKS

The only EDS books out there that I know of are all technological books. There was one book I read, and it had a section on "patient experience." I was extremely excited that somewhere out in the world there was something highlighting and, in a way, exposing how we are forced to live. They got the general patient experience correct but they blew it on helping the EDS community with anything other than medical information.

The section on what they labeled as the "patient

experience" was about half of one page in a 600-page book. The "patient experience" was a huge let down. Our lives don't just sum up in a paragraph; they can't even be summed up in this book! People need to know that there is an entire community that is disabled and abused by medical negligence. We are even more abused when we try to get the medical help we need but doctors so easily dismiss us. This only adds to the isolating experience that can be EDS.

## EDS SUPPORT GROUPS

Nothing is better than finding someone who understands you and "gets it." Not only does it validate your whole life, but you can also make friends who will be a big support going forward. You don't have to go through anything else alone. We have a lot of symptoms and problems with areas that are extremely personal. These are things that I wouldn't normally share with anyone that I wasn't living with. All of this binds the whole EDS community together and you will feel a lot closer to someone battling EDS than you are with anyone in your family. It seems overwhelming to go up to a stranger but it's not. Everyone you find in these groups is very welcoming and you will get all your questions answered; no one is ignored. I had a question and went to a POTS / dysautonomia group since the best information you learn will be from others and their experiences. While I was on the group page, I saw a post asking if there was anyone who was available to talk. I really had no idea how much I would get when I answered her post.

That was two years ago, and Kim became my best friend from the first phone call. She's always there for me, whether it's just to chat or to have an ugly cry. I'm stronger and my mental health is lighter because of her. I've seen her get knocked down and she always fights back. Even if she doesn't feel brave, she certainly does show it. Through her, I met Christie; we are so similar we might as well have been twins since birth. Seeing her fight back, her strength, and her bravery have made me a better person. If it were not for these two ladies, I know how hard and dark my life would be.

There are several Facebook groups for all disorders – EDS, POTS, MCAS, dysautonomia – both local and worldwide. The EDS society website has so many resources and places to meet other people. There's National Organization for Rare Disorders (NORD), and Dysautonomia International has great resources and a way to look for specialists in your area. I have no doubt that there are many more out there. These places are a great place to start. Here you will also learn about other groups.

## IT IS LIFE AND DEATH

The hardest part of this disorder is not being believed and therefore medically neglected. Even if it takes years, it's important to keep looking until you find the doctors who are right for you. We all need to make sure we find those people because one day, you may need them to save your life. It's no secret that EDS patients will spend a lot of time in hospitals and doctors' offices. We are in life and death situations.

Just as I was beginning to feel better, lose some water weight, and have improved cognitive function, I came down with pneumonia for the first time.

I have always had problems with respiratory illness, viruses, and most often bronchitis. After recovering from each instance, I had a few weeks with a post infections cough. So, it was no surprise to me when I ended up with pneumonia. At first, it was very painful. Each breath felt like I was inhaling icicles and it got increasingly harder and harder to breathe. I felt like I wasn't getting any air. I was seen at urgent care and sent home on antibiotics. Only two days later, I couldn't breathe, my fever was up to 102°, and my lips had a bluish hue. I was miserable;

to top it off, I was still passing out several times a day. I blacked out and went straight down. I remember hearing myself scream, though it sounded like it was coming from very far away.

When I came to, my right ankle that I'd had surgery on a year ago was badly sprained. Since I was getting worse instead of getting better, we packed up and headed back to the hospital less than three months after being discharged. I was admitted and diagnosed with pneumonia, a pulmonary embolism, sepsis, and I had fractured my ankle – could it really get any worse?

The following morning, I met PA Moyes who gave me a thoracentesis (use of a syringe to extract fluid from the body). I was hospitalized again, and new CT-scans showed a large amount of fluid in my chest cavity. A small tube was inserted into my back coming out of my side with a bulb reservoir to allow the fluid to drain. After a day, two liters of fluid were drained from the space between my lungs and inside wall of my chest cavity. Another CT-scan was done to determine how much if any fluid remained. It was determined that there was still enough fluid to keep draining. For the next few hours, there was little drainage, and I was given tissue plasminogen activator (TPA) to "unclog" my drain. Not much later, I noticed that the drain was full of dark thick blood.

The color indicated that it wasn't an active bleed but blood that was old and stored in my body.

My nurse that evening was one of my best. She took pictures and contacted the on-call doctor every few hours. By midnight, she felt that I didn't look well. I was pale and my skin had a grey tint. I was given a blood transfusion of one unit and by 1 am, my nurse decided that she wasn't going to wait for the doctor to get back to her and sent me to the ICU. She told me that I didn't look right, and her intuition was dead on.

Once in the ICU, I was given another unit of blood and one bag of fresh frozen plasma (FFP). My husband had also come to the hospital. I was having a difficult time coping with being returned to the ICU just days after I had been moved to the tele unit from the ICU. My husband left about an hour later. Looking back, I was glad he left when he did. He already had so much on his plate with two jobs and a teen in high school who did remote learning to prevent contact with other students. Since COVID was everywhere, my son didn't want to bring home a virus that would kill me.

After my husband left, it was calm and quiet for about three hours. I hadn't been able to sleep. I was anxious about being back in the ICU. Around 4 am, I heard an announcement over the PA that an emergency response team was needed for a mass blood transfusion in my room. I had no idea when or what happened since I didn't see anyone since my husband had left. My confusion turned to realizing that what was going to happen would take place in my room, and I felt adrenaline searing through me. I know that if it was announced over the PA that it wasn't just an emergency – it was extremely urgent. Chaos ensued with twenty to thirty nurses appearing, one of them rolling in a huge machine I had never seen before. I was panicking and was very close to losing it, the next few seconds felt like forever and in slow motion.

I scanned the room looking for a nurse I knew. I had sudden calmness when I saw Gabe, a phlebotomy technician who I knew quite well. Somehow seeing him calmed me down. I experienced a feeling of security. I watched everyone buzzing around the room, blood and plasma being loaded into the machine – a total of eight units of blood and four units of FFP.

Later I learned that the human body holds about ten units of blood or about five liters of blood. I had a PICC line because my veins are very hard to get an IV started. The nurse connected me to the line and the machine started to drain the blood. Gabe had to take blood after each unit to test to track how much I needed. A doctor came in just as the nurse announced that my PICC line wasn't working fast enough and that I was still bleeding out faster than they were able to push the blood in. Her voice is forever seared into my brain. I was still bleeding out. Gabe didn't hesitate; he quickly grabbed what he needed, and he was able to start another IV line in a large vein on the first attempt. Under all that pressure and urgency, Gabe was able to do what was needed and to do it quickly.

My IV was hooked up and working and I was no longer bleeding out faster than the blood was being replaced. Everyone stopped what they were doing and watched the machine work. They were all very still except for the doctor; as I watched her, it hit me just how dire my situation was and how close I came to dying. Doctor Janowski had one arm wrapped tightly around her waist and her other hand was by her mouth as she was biting her nails, while she paced back and forth quickly. She looked as if she was going to vomit. I was terrified, paralyzed, when I realized that the doctor wasn't sure if I was going to live or die. I can't even begin to describe how I felt as soon as that thought crossed my mind. There was a real chance that I wouldn't survive.

Everything happened so fast, and the machine only took a few minutes to push all the fluid. I saw the relief on everyone's faces. It was like the whole room let out a long sigh. I was in a stupor, couldn't think, couldn't breathe. The emergency was over, and I was going to make it. Everyone trickled out and I was soon alone again. I couldn't believe I had survived. It was surreal.

By 7 am, a nurse, Brandee, who sent me back to the ICU, showed up. I'm sure the intensity of the situation laid heavy on her. While we were talking, she told me how she wasn't sure I lived. Since she sent me to the ICU, she knew what room I was in and had heard the PA announcement. She was extremely busy that night. She had a lot of patients and was the charge nurse for the shift. She told me that all night all she did was chart the amount of blood she had emptied from my drain. She was on autopilot, so it didn't occur to her to add the total amount of blood. I remember so clearly the look on her face as she talked – the complete disbelief that I was still alive. She had charted almost six liters. I began crying knowing that a miracle had occurred. If it weren't for Brandee sending me back to the ICU, for Rachel who loaded and operated the machine, and for Gabe being able to start an IV on the first try, I would not be alive. These people saved my life, and I will never forget what they did for me. There are just no words to convey my gratitude.

Hearing about something, like a life or death situation, is very different than experiencing it and being there in the moment while it happens.

Scars show us where we've been, but they don't dictate our future.

# 12

# BE YOUR OWN ADVOCATE

## DOCTOR EGO

For some reason, our society has placed emphasis on prestige and respect. To doctors, there's no question that we all assume that when we got to a doctor, we're going to be taken care of. It's largely believed that the reason someone goes into the medical field is to help people. Unfortunately, most doctors become doctors for themselves. The special status, respect, and money appeals to many of them. You don't have to spend a lot of time with a doctor to figure out if they are truly there for you and your health, or if they are there for their own benefit.

For many, many years, I dreaded doctor appointments, especially if it was with a new doctor. It was so anxiety-inducing that I'd literally be shaking. I was always armored and ready for a fight. Every time, I was prepared to argue and prove to them that what I said was really what was happening. I was setting myself up and automatically assuming that they weren't going to believe me. There was always such a pressure on myself to get the care I needed. This was emotionally and physically exhausting. I usually spent one or two days in bed to recover from the one appointment.

You are the expert on you and most medical professionals have never heard of EDS. Being the expert and having to manage our medical needs ourselves means we really need to be the one in control. Doctor appointments can be devastating when the doctor is standing between you and the treatment you need. There is the potential that we aren't believed. We may be told that it's all in our heads – that we are liars and fakers – and we leave the appointment even further away from getting what we need.

What we need to do is take the power back; be the one in control of which doctors you see. When we go out to restaurants, there is an expectation of service, kind, attentive servers, prompt food that's cooked correctly without strands of strangers' hair. There are a lot of reasons why we would complain or speak to a manger to get the meal remade and most likely get comped from your bill if you found something wrong with it. If we are confident in returning our meal, then why aren't we able to make sure that the service we are paying for in medical care is up to our expectations? Why aren't doctors held to the same type of standards of respect in patient care? When a doctor doesn't give you any respect as a human being, then why don't we do anything about it when we are unsatisfied? When a doctor doesn't give you any respect as a human being, then why aren't we bringing attention to it? Why don't doctors get held accountable for unacceptable behavior?

Our health is extremely dependent on medical support just to keep us alive. There are more times when we are treated with blatant disrespect than there are appointments when we are treated with basic respect. We aren't listened to, taken seriously, or given the benefit of the doubt. What's more frustrating is hearing about real instances when someone with Munchausen's has gotten an abundance of health care, even though they don't have proof of the diagnosis they claim to have. Doctors have tried to convince me that nothing is wrong with me, even with my diagnosis from a well-respected geneticist and physical lab results showing the cause of my EDS.

We can take the control back by adjusting our

perspective and thought process by prioritizing getting the right doctor to treat you. The right doctor for you is out there; don't stop looking until you find them! Unfortunately, there is no way to tell how long it will take, how many doctors you will meet with, or how many times you will be told you're crazy, lying, or faking before you find the doctor for you. It took me almost twenty years before I found my first doctor. Since it took almost nineteen years from the time I become symptomatic to being diagnosed, I ended up needing a team of doctors to work together.

Now four years after diagnosis, I have most of my own team of doctors. My team has a cardiologist, neurologist, autoimmune doctor, pain doctor, functional medicine and dysautonomia doctor, a gastroenterologist, PCP, nutritionist, psychiatrist, psychologist, a case manager for my health insurance, and a hospital that knows me well. It was difficult to go to all my appointments physically and the doctors and nurses that only added to the PTSD were extremely grueling. However, I would do it all over again, no matter how hard it was. Now I have what I need for the next forty to fifty years. I will also never have to worry that my children will have to experience the same degrading treatment because they will see my team of doctors.

## IT'S AN INTERVIEW

Since I was tired of having no control over my life and tired of being angry after every failed attempt at finding the right doctor for me, I began by turning the tables. I was tired of not having any kind of say, especially if the doctor made up their mind about me before they even talked to me. I decided to look at it like I'm filling a position; the doctor I'm looking for must meet or exceed my expectations. There are a lot of doctors out there, and we want one who doesn't mind doing some research or homework so they can give you the best care.

Decide first what it is you are looking for in a doctor. This may change depending on what type of doctor you need. Once you know what you're looking for, you should search for all the doctors who are in your area or doctors who your insurance will cover. It's a personal choice how much research you do on the doctors. Generally, I only see one doctor at each clinic. I've had experiences where one doctor refused to treat me because I didn't do everything she wanted me to do. I did everything she asked for six months before I got any treatment. At that time, I was on pain medication to help me eat, but she wanted me to stop taking the pain medication. I told her that stopping the meds was the only thing I *couldn't* do because that was the only option I had so I could eat. If I didn't have pain meds to eat, I had two options, one was to end up in the ER every night after only eating dinner or starve. My kids were old enough to understand that I couldn't eat, and no one would help me. It was gut-wrenching to see them watch me basically starving to death. I was still trying to find a GI doctor who would help me eat. Years later, after being diagnosed, I needed to go back on pain management. Still, no doctor in that practice would treat me.

After that experience, I would only see one doctor at every clinic. Everyone knows that you must go on several interviews before landing a job. So, I planned on seeing around five doctors per specialty. It's only considered doctor shopping when you go to multiple doctors and fill multiple prescriptions for controlled substances. That gave the power and decisions to me. How I was treated determined which doctors I chose. Now I'm free of the pressures and stress of hoping that the doctor believes me. They either did or didn't. They either met my expectations and needs, or they didn't. There's no way you should have to settle, especially for a doctor who doesn't give you any respect. It's okay to know what you want and it's okay that you get it.

Doctor's behavior never seems to be in question.

It seems as if they can treat us with such blatant rudeness and disrespect and there isn't anything we can do about it. They know that they can control our lives by being in complete control of our healthcare. I can't stand the doctors who use their position to control others like it's a game. Except for us, it isn't a game, and it isn't general healthcare; for us, it is a battle for life or death. That may sound dramatic, but for us, it is that serious.

More of us need to be comfortable with just walking away. More of us should be leaving reviews on doctors who don't treat you well. More of us need to have our stories heard. That's the only way we can let others know what type of doctor they are going to. It may take longer to get what you need, but it's still faster to keep looking and meeting with new doctors who will help you, than it will be to get the proper healthcare from a doctor who enjoys watching you be defeated when they deny you what you need. We need to be our own advocate, as no one will do it for you. Don't be afraid to ask for what you need and don't settle for a doctor who doesn't treat you like an equal.

## 13

## HAPPINESS CAN BE POSSIBLE

No one plans on getting sick. It's not a valid career or lifestyle choice. It would be great if I had the option to take my body to the store for a refund because I'm not satisfied with it; after all, I got a defective one. Becoming disabled has been the single most difficult challenge of my life. For so many years, my family and I all talked about how when I got better, how after I healed, then we could do X, Y, and Z. I felt like I was in a world of limbo with no actual progression in life. I was always kind of stuck in a continuous loop of doctors' offices and hospitals.

On top of this, my family had to handle the doctor bills that we couldn't pay off. There was a time we had as many as twenty-one different doctors, surgeons, dentists, anesthesiologists, and hospitals sending us bills. When I became symptomatic, we had to pay these medical bills on top of our monthly health insurance cost. We met our max out of pocket expenses that at one time had been $9,000. We've reached our out-of-pocket costs by March for the last twenty-three years. Living with financial strain was difficult at first, because we had gotten married young, and our daughter was born only two weeks prior to our one-year anniversary. Then I got smart and very creative with our situation. We couldn't afford to take trips to Disneyland or long camping trips, but we couldn't live life only being angry and frustrated.

We eventually decided that living this way, and with the anger and frustration, was too hard. We had to be okay with the constant appointments, we had to be okay with the spontaneous ER visits and hospitalizations.

All we could do is control how we acted in response to what was happening. So, we lived our lives and made plans and scheduled stuff, but we also knew that something was going to happen. We just didn't know what it was, and we didn't know when it was coming; we just knew that it was. Any changes to plans weren't a big deal because we went into everything knowing something could happen. Some things worked out and other things got rescheduled. We didn't let the negative emotions control us and the only way to learn that is with experience.

I've been angry, I've been frustrated, and I've been downright miserable. Then when I turned to the people who could help me get a better life, I ended up feeling like it wasn't a battle with my health, but a battle against doctors, specialists, and all clinical staff, especially when health concerns are quickly brushed under the rug and dropped.

Going through the surgery, recovery, and life was a big struggle, and an adjustment, but with our new outlook, it is so much better. We are able to wake up happy and in a great mood. As time went on, my medical condition became our new normal. The biggest secret to a happy life is your perspective. As your medical history continues to get longer, it is important for you to feel each experience. You can't just lock them all away before you've had some time to feel all your emotions for each experience. If you don't allow yourself to feel those experiences, they will pile up, and one day, you'll be in the middle of an upsetting moment. Instead of just going through your emotions for this one experience, all the ones you locked away will hit you. It is a PTSD response, and it will be extremely sudden, drowning you emotionally. With EDS, you know that something else is always around the corner. It is much healthier to allow yourself to be upset and grieve over these experiences one at a time. Working through your emotional state a little at a time helps to avoid a mental dam from being forced open. Take the time you need to work through it and then let it go. Be cautious that you don't dwell for too long on any of these experiences. Dwelling on one for too long, it turns into a slippery slope and it's far too easy to slide down and become hyper-focused on it, and then it's even harder to learn to cope with and be ready for what's next, since you know it's coming.

## HUMOR HELPS

After a decade of trying to figure out my care team, I knew that change had to come from me. I wanted some peace and a sense in normalcy back in my life. I could not keep having thirty to forty meltdowns a year. I made the choice to add happiness and joy back in my life. I was done being angry, upset, and hurt by everything. I wasn't the type of person who was constantly crying and the only way I knew to avoid that was to constantly be laughing. Humor is one of my coping techniques and I've used it for as long as I can remember, and it's also a good idea for you to learn a few ways to cope and find a coping mechanism that works for you.

As I've gotten older, and my EDS has greatly increased, and my humor has gotten a little bit darker but is never self-deprecating. After all, putting yourself down will only make you feel worse. It implies that somehow you are to blame for your health.

The last time I stayed in my hospital, I needed an X-ray of my chest. Everyone knows that the techs are not able to disclose things they may notice on the X-ray, only a doctor can give a diagnosis. As I was getting into place, I asked the tech if she could do something for me. She gave me a look that said, "Is this lady really asking me for something that I'm not allowed to do?"

I then told her I knew she was going to look at all my hardware, so I'd really appreciate it if she could tell me what bionic level I had reached. I always love it when any staff plays along with me. When she came back, she told me I had quite the impressive make up and that she would have to say that I had attained a bionic level of 8. I then asked her excitedly what level I would be when I'd have a titanium plate put on my ankle to help keep my tibia bone to stay in place a lot closer to my foot bones. She said I'd be a level 9 for sure, and that level 10 was the highest. Then I got really excited about it and I told her that when I reach level 10 then I would become the bionic queen.

I mentioned one of my goals in my professional patient career is making staff laugh. My second goal is that when I am healthy enough to, I am going to storm the school board. Between all six teenagers who I cared for and my two high schoolers, not once during those years, nor during my own, did I ever attend a career day that had included my profession! I mean, a little heads up of what to expect would have been extremely useful!

## ACCEPT YOUR LIMITATIONS

Another thing that helps manage EDS is accepting and working with your own limitations.

For so long, I was holding myself to doing things the way I had always done them. The longer I tried to keep up with my life, the sicker I got. I needed to find a way of taking care of myself the way I needed to be taken care of. In order to do that, we need to approach everything with our limitations. Once you stick to living within your limitations, you'll be less fatigued, experience less pain, and will dial down your stress. It's the same as doing your budget. When you live outside of your means, you have more debts, restrictions on your money, and have less money to spend freely. When you do live within your means, you have less debts, need less money for bills, are not collecting interest fees, and you're just much freer to spend your money as you choose.

Our bodies aren't as strong as they used to be, and we need so much more rest to function. Learning what your specific limitations are allows you to know what to look for to help you for the better. An example is when we go places where there is a lot of walking, say to the zoo. I could walk the entire place but then I know I'll be down and in bed for several days after. So, to still enjoy the zoo and not have to recover afterwards, I choose to use my wheelchair and have someone push me. It doesn't take near as much energy and effort. I don't have to clear a week for two days before and two days after to rest to be able to enjoy a trip to the zoo.

Especially right after I was diagnosed, I wanted more answers, more help, and I didn't want to wait. I needed to see a cardiologist, neurologist, GI, orthopedic, pulmonologist, I also needed a sleep study among so many other things. I'd called and made appointments with everyone I needed. As time went on, all the appointments, tests, treatments, and follow up appointments were really taking a toll on me physically. I needed more and more days of rest before and more days of recovery afterward. I couldn't keep up with the amount of time I was spending in doctors' offices.

Eventually, I was forced to slow down. As hard as I tried, I just couldn't keep my eyes open; my body just gave up. When I was able to stay awake, I was still very weak and basically in recovery mode. I needed a lot of rest, and I slowly was able to function. As everyone with chronic illness knows, we can't physically keep up with regular life. Our bodies physically can't function at 100 percent.

It's not enough to just know your limitations; with EDS, you need to live your best life *within* your limitations. I knew I couldn't keep seeing so many doctors so close together. Unfortunately, we have to decide which issues we can work on, and which ones need to take the back burner. I functioned so much better by just cutting back on the number of appointments I went to each week. I decided to focus on two or three things at a time, especially with new doctors because this is when you have to get the most testing done.

This takes a lot of acceptance on our part. Acceptance that we have limitations, acceptance that we can't fix everything right away, and acceptance that something is always coming, we just don't know when it will show up and what it's going to be. With acceptance, we can live a more stress-free life and it's much less upsetting when you have to rearrange your schedule to accommodate how you are feeling.

Additionally, I'm constantly changing the braces I need; most of the time, there is no injury, but I use the braces for support. I also own a walker and wheelchair, and it's important to use walking aids if we need them. I'm always in and out of my wheelchair. I use it on days I am really dizzy and experiencing syncope. It's always nice to use it if we are going somewhere and I have to walk a lot. It helps save energy and allows me to be out all day without overdoing it, so I can enjoy being out and not worry about the consequences. I don't care if people see me walk right over to my chair. Just because I don't use it 100 percent of the time doesn't mean I don't need it. Other people might not understand our limitations, but we do, and this is one way we can take care of ourselves.

## CHOOSE TO SHIFT YOUR MINDSET

The winter of 2010 was a difficult one for our family. I was having frequent breakouts of shingles on my forehead and nose every few weeks. It's one of the most painful things that you can get. On top of that, I had bronchitis constantly. After four months, I had gone to the hospital for a CT scan. A few days later, I had the call I was waiting for to hear the results. I got results, just not the kind I was expecting. I was waiting for a call about my lungs, but instead I got a call about my heart. My heart was grossly enlarged, and I was told that I needed to go to the State University Hospital to see a cardiologist and it was urgent. I was in the cardiologist's office in one to two weeks, when normally an appointment takes five months to get.

While there, I was told I had a little different sound to my heartbeat and was setup for a test where they went up my femoral artery to get a video of my heart while they put a dye in to see the blood as it pumped through my heart. Within a few days, I was told to get my blood taken and start a blood-thinning medication. After starting the medication, I had to go and get my blood taken every few days to make sure my body's level of medication remained in a certain range.

After two weeks, I got a call and was told that the head cardiologist had been on vacation and there had been a miscommunication between the doctors, and what they thought was wrong wasn't the problem. I was told to stop the medication immediately and I was scheduled to see the doctor who was at the cardiology clinic. I wasn't sure what was going on, but I was so worried that I couldn't sleep that night.

The next day, I was back at the hospital to see the head cardiologist. There were a few things I had repeating in my head and none of them were good. Even though I've been told over and over that I'm unique, I was not expecting just how few people had the same issues I did.

I was told that I had a birth defect on my heart, and it should have been caught when I was little. It would have been a very little, non-invasive procedure, but since it had grown over the last thirty years, the fix wasn't so easy. In fact, my best chance was open heart surgery to fix it. I was only the doctor's fifth case he'd ever seen in his thirty-plus years of practicing. My surgeon had been doing heart surgeries for ten years but there was no other doctor who had ever seen this defect before.

I was told it would be a very difficult three-month recovery. At the time, I had no idea about EDS, and my recovery would actually turn out to be six months minimum. During the procedure, I had my heart stopped for about eight hours and a machine was pumping my blood and breathing for me, because the surgeons couldn't work on it while it was actively pumping.

I talked about my heart surgery before, but it is worth repeating, because this was our first time as a family when my life was on the line. We'd been struggling with feelings of anger, frustration, hurt, and confusion. The constant questions of "why?" loomed over us.

When we first heard that there was a potential problem with my heart, we begin talking about all the negative feelings we've been living with and how hard it was. I didn't want my last moment with my family to be with such heaviness in a downright depressing state.

Being faced with such a real possibility of me dying, we decided then and there that we couldn't wait for life to be easy before we decided to be happy again. Ever since we had met, our relationship was full of fun and so much laughter. If I did pass away early, I wanted my husband and children to remember me with happiness, love, and laughter. I did not want them to think of me as just merely surviving. We changed our perspective. Everything we did going forward, we got through with humor and fun. Our lives were no longer grave. We got through one day at a time and took on a couple health problems that we felt were the ones that were most pressing. ER visits became a "date night." We had to get sitter for the kids, and we were able to talk and watch a movie while we waited for test results. There was always a way to turn the situation around and find that silver lining.

The day before my surgery, I had to come to the

hospital for some tests and collect some blood for replacement. I had an ultrasound, X-ray, more blood work, and even an EKG. An EKG tests the electrical signals of your heart to look for any problems. Depending on the need for the test, you'll get electrodes placed on your head, face, heart, and all your limbs. It's taken while you are laying down and only lasts for ten seconds. It is not painful, and you don't feel anything.

When the EKG tech arrived, it was a very young man, and he had the biggest smile on his face. The second he came into the room, he started to speak quite fast, gushing about how this test was his first test solo. He didn't ask if I'd ever had an EKG and kept chatting to keep the feeling in the room to be upbeat and have great bedside manner. His name was Rocky, and he reassured me that this test only hurts a little bit and was also very quick. It took ten times longer to attach all the electrodes. Now, what he didn't know was that I was the one person he shouldn't be telling all this to. I just calmly agreed with him and made no effort to just tell him that I'd had many, many EKG tests over the last decade, and I knew exactly what to expect.

I watched him very closely to get my timing on target. I waited for him to hit the start button, and as soon as he hit the button, I started acting like I was having a seizure and didn't hold back on the volume of the sounds I made. He instantly stopped the test, and he had the most terrified look on his face, stumbling on every word. He kept assuring himself that he'd been told that it wouldn't hurt at all and asked me repeatedly if I was alright. He switched from exhilarated to pure panic. I was absolutely hysterically laughing, and my husband cracked up. Then the nurse who'd been in the room since we got there was laughing so hard that she had to catch herself from falling off of her chair. I still remember her saying that I had just made her entire week and that my timing was perfect. I can only imagine what the response was from their co-workers and later that night their significant others, as they must have said, "You'll never believe what happened to me..."

After we adopted and normalized our new perspective, life became so much easier to get through. I was waking up all the time in a great mood. We use humor to cope with all the situations we find ourselves in and it really releases the pressure.

Our family really had to come together in order for us to pull through. At the time, I felt very fortunate. The surgeon told me that had I not found out or if I waited to have surgery, sometime in the next three to five years I would have no longer had the option to fix my heart. After that, my life would have ended in my fifties or sixties from over-working my heart. I was so close to missing the time I had to save my own life.

Happiness truly is a choice, but it's not instantaneous and it takes a lot of work. There is a process, and it starts with deciding if being happy is a goal that you want to achieve and that you will do the work to achieve.

Every so often, I have a complete mental meltdown. It's okay to not be okay. It's important to allow yourself time to feel sad and have a safe place to have a good, ugly cry. It's been proven that this has a positive effect on your body. It doesn't mean that we pitch a tent and have an extended pity party. Do take the time you need, and then get up, brush yourself off, and get ready for the next battle.

We all must remember that it's not just crying that our body needs. What's equally important to your healing is to have a "tears running down your leg" laugh. It's going to be more cathartic than you expected.

We are all warriors in our own right.

# ACKNOWLEDGMENTS

For all those who have supported me, cared for me, and have made this dream possible, I want to say thank you. More importantly, I want to express my everlasting love and overwhelming gratitude to all. Each one of you has touched my life, guided me, kept me alive, and has shown me the ultimate sacrifice of service. I may not remember everyone's name, but I will never forget all those who touched my life.

First, I want to thank Dr. Angela Lauria, Madeline Kosten, and the entire Difference Press team for allowing me the time to heal, and all their hard work in making this book become a reality.

My husband, Gerad, for working so hard and being by my side from day one. My children, Morgan and Logan, who have always given me a reason to smile, laugh, and find fun in everything. My girls, Fei, Eva, Rachi, and Miri, who are the most precious additions to my family. Also, my real father, who shows me every day how much he loves me.

My Papa Joe and Great Grammy, both of whom showed the ultimate kindness from the day we met.

Frank and Nancy Webb, who were the parents I needed and taught me what family really means. Susan Pullan, who after almost thirty years, continues to be a true sister. The entire Cruz crew for supporting me and my husband.

Darci Weis, whose friendship has literally saved my life. Rebecca Farell, who is the sweetest lady I know and someone who knows how to really get a party started. Jenny Anderson, for being my sister and friend. Kim Bluuemel and Tiffany Woolsy, who are the best neighbors and are always checking in on me. Debbie Wood, who always has a bowl of homemade chili for me.

Kim Ryberg, who showed me what a real best friend is. Christie Hanchey, who is literally my twin and medication. These two are the perfect example of people living their best life among the worst circumstances and continue to thrive against the worst odds imaginable.

Myranda Wilson-Arave, for surviving the worst together and teaching me a lot about EDS. Kendra Stevenson, someone who has shown my unwavering courage. Alicia Pearman, who continues to be a shoulder to lean on. Shelly Gee, for being brave enough to speak to a stranger and introduce to me to EDS. Without her I probably still wouldn't know what I was dealing with, and she has changed the direction of my whole life for the best.

Dr. Abu Aba Care, who is the only doctor in my twenty-three years' experience who was present during a CT scan so he could get the results without delay. He is the reason I have a hospital I can go to and don't have to be afraid. As those who rack up frequent flier miles from a hospital bed know, this is the greatest gift I could ever receive. Along with his team of doctors, Dr. Adepu, Dr. Janowski, PA Strong, and PA Moyes, who make this hospital so special.

My whole doctor team, past and present, Dr. Morrill, Dr. Richards and Kelcee, Dr. Tueller, Dr. Adams, Dr. Garg, Dr. Whittleder, Dr. Law, Dr. Gonzales, Dr. Bilstrom, Marysa, Dr. Hancock, Dr. Hansen, Dr. Kinikini, Dr Kessler and the entire wound clinic, for making my life still possible.

The amazing ladies who I had the absolute pleasure of meeting and working with, Tami Stacklehouse and Annette Jones, both of whom taught me that merely surviving isn't living and that I am much more capable than I ever imagined.

Nurses and all hospital staff, including PT, radiology techs, housekeepers, and cafeteria workers who have ever cared for me, Hannah, Brandee, Kylie, Leni, Theresa, Lisa, Melissa, Harley, Rachel, Gabe, Somner, Aurora, Noelia, Derrick, Kelsey, Hogan,

Peggy, Leslie, Jess, Ian, Donna, Kathy, Morgan, Charolette, and Megan. As much as it pains me to not list everyone by name, the sheer amount of people who cared for me and the consequences of severe infection, sepsis, almost dying on multiple occasions, and some of the most beautiful ethnic names I've ever heard and for the life of me can't spell, I will never be able to do so. I will also ever be in their debt, thank you all for giving me not only the will to live, but also the means to live!

## ABOUT THE AUTHOR

Julee Cruz has been a professional patient for the past twenty-three years after becoming symptomatic with Ehlers-Danlos Syndrome (EDS) with her first pregnancy at the age of twenty. Because of her condition, she had to drop out of college before getting her bachelor's degree, was unable to have more children, was bedridden for three years, and has had countless painful surgeries.

As many with chronic conditions can relate, Julee has seen hundreds of doctors, spent two-and-a-half years in the hospital, has had ten near-death experiences, and has been dropped as a patient for being "crazy" and "too complicated."

However, as she dealt with her condition, Julee also found great doctors and wonderful friends. Julee doesn't apologize for having EDS and is determined to keep her experiences and circumstances from changing her. She aims to help others who are living with EDS and chronic conditions to have a better quality of life and to make a difference in the world.

Julee lives with her husband, children, and father in Syracuse, Utah.

# ABOUT DIFFERENCE PRESS

Difference Press is the publishing arm of The Author Incubator, an Inc. 500 award-winning company that helps business owners and executives grow their brand, establish thought leadership, and get customers, clients, and highly-paid speaking opportunities, through writing and publishing books.

While traditional publishers require that you already have a large following to guarantee they make money from sales to your existing list, our approach is focused on using a book to grow your following – even if you currently don't have a following. This is why we charge an up-front fee but never take a percentage of revenue you earn from your book.

### ☞ MORE THAN A COACH. MORE THAN A PUBLISHER. ✍

We work intimately and personally with each of our authors to develop a revenue-generating strategy for the book. By using a Lean Startup style methodology, we guarantee the book's success before we even start writing. We provide all the technical support authors need with editing, design, marketing, and publishing, the emotional support you would get from a book coach to help you manage anxiety and time constraints, and we serve as a strategic thought partner engineering the book for success.

The Author Incubator has helped almost 2,000 entrepreneurs write, publish, and promote their non-fiction books. Our authors have used their books to gain international media exposure, build a brand and marketing following, get lucrative speaking engagements, raise awareness of their product or service, and attract clients and customers.

### ☞ ARE YOU READY TO WRITE A BOOK? ✍

As a client, we will work with you to make sure your book gets done right and that it gets done quickly. The Author Incubator provides one-stop for strategic book consultation, author coaching to manage writer's block and anxiety, full-service professional editing, design, and self-publishing services, and book marketing and launch campaigns. We sell this as one package so our clients are not slowed down with contradictory advice. We have a 99 percent success rate with nearly all of our clients completing their books, publishing them, and reaching bestseller status upon launch.

☞ **APPLY NOW AND BE OUR NEXT SUCCESS STORY** ✍

To find out if there is a significant ROI for you to write a book, get on our calendar by completing an application at www.TheAuthorIncubator.com/apply.

## OTHER BOOKS BY DIFFERENCE PRESS

*Fundraising without Burnout: Radically Reimagining Philanthropy to Transform Your Impact* by Radha Friedman

*Always Bring Your Sunglasses: And Other Stories from a Life of Sensory and Social Invalidation* by Becca Lory Hector

*Art of the Heart: The Doctor-Patient Partnership* by Jay H. Kleiman, MD

*Living Intentionally after Loss: 8 Steps to Reclaiming Your Passion and Purpose* by Maya Manseau

*Breakthrough to Entrepreneurial Brilliance: Shatter the Invisible Barrier Holding Your Business Back* by Alana Mills

*Is This a Cult?: Confronting the Line between Transformation and Exploitation* by Anne. L. Peterson

*Founder to Exit: A CFO's Blueprint for Small Business Owners* by Pam Prior

*Prove Them Wrong: One Immigrant's 10-Year Journey from Bankrupt to Millionaire* by Héctor E. Quiroga, J.D.

*A Second Wind after Loss: A Guide to Health and Renewed Purpose for the Grieving Heart* by Denise Sherman

# GIFT FOR READER

I want to share my deep appreciation for anyone else who struggles with EDS and other chronic illnesses. As my gift to you, I'd also like to share the team who supports me.

    Christie has an email, a podcast, and a website. Christie has a lot of experience with fascia and fascia remodeling. You can reach her at AWOLZebra@gmail.com and AWOLZebra.com is also launching on June 20$^{th}$, 2024!

    I have been a guest on the podcast several times, so if you want to know more and want to listen to us laugh at ourselves it's at AWOLZebra.

    If you'd like to get in touch with me, you can email me at fragilehandlewithcareeds@gmail.com.

    We all will do our best to answer as quickly as possible since we have to live with EDS too.

www.ingramcontent.com/pod-product-compliance
Lightning Source LLC
Chambersburg PA
CBHW052150220526
45471CB00004B/1616